Springer Theses

Recognizing Outstanding Ph.D. Research

For further volumes:
http://www.springer.com/series/8790

Aims and Scope

The series "Springer Theses" brings together a selection of the very best Ph.D. theses from around the world and across the physical sciences. Nominated and endorsed by two recognized specialists, each published volume has been selected for its scientific excellence and the high impact of its contents for the pertinent field of research. For greater accessibility to non-specialists, the published versions include an extended introduction, as well as a foreword by the student's supervisor explaining the special relevance of the work for the field. As a whole, the series will provide a valuable resource both for newcomers to the research fields described, and for other scientists seeking detailed background information on special questions. Finally, it provides an accredited documentation of the valuable contributions made by today's younger generation of scientists.

Theses are accepted into the series by invited nomination only and must fulfill all of the following criteria

- They must be written in good English.
- The topic should fall within the confines of Chemistry, Physics, Earth Sciences, and related interdisciplinary fields such as Materials, Nanoscience, Chemical Engineering, Complex Systems, and Biophysics.
- The work reported in the thesis must represent a significant scientific advance.
- If the thesis includes previously published material, permission to reproduce this must be gained from the respective copyright holder.
- They must have been examined and passed during the 12 months prior to nomination.
- Each thesis should include a foreword by the supervisor outlining the significance of its content.
- The theses should have a clearly defined structure including an introduction accessible to scientists not expert in that particular field.

Luca Anghinolfi

Self-Organized Arrays of Gold Nanoparticles

Morphology and Plasmonic Properties

Doctoral Thesis accepted by
the University of Genoa, Italy

 Springer

Author
Dr. Luca Anghinolfi
University of Genoa
Italy

Supervisor
Dr. Francesco Bisio
University of Genoa
Italy

ISSN 2190-5053   ISSN 2190-5061   (electronic)
ISBN 978-3-642-44082-3   ISBN 978-3-642-30496-5 (eBook)
DOI 10.1007/978-3-642-30496-5
Springer Heidelberg New York Dordrecht London

Printed on acid-free paper

Springer is part of Springer Science+Business Media (www.springer.com)

Supervisor's Foreword

Metals largely owe their characteristic optical response, like the almost unitary long wavelength reflectivity, to the interaction of electromagnetic radiation with their free electrons. As soon as a metal is manufactured in the form of a nano-metric-sized particle, however, the physical confinement that the particle's boundary exerts on the free electron gas leads to a dramatic change in its optical response. Metallic nanoparticles with sufficiently low losses, under the influence of an electromagnetic field, can exhibit in fact a resonant enhancement of their polarizability that is completely absent in bulk counterparts and leads to a highly peculiar optical response. Such a behavior, that goes under the name of localized surface plasmon resonance (LSPR), leads to a strong absorption of radiation by the particles, and to extremely strong enhancements of the incident field on a typical length scale of just a few nanometers away from the particle surface. These exciting features have disclosed the possibility of designing materials with tailor-made optical response and most importantly have provided a method for harnessing the propagation of the electromagnetic radiation on subwavelength scales, with heavy applicative implications.

In his thesis, Luca Anghinolfi has addressed the potential of applying a totally self-organization-based approach to the fabrication of spatially extended 2-dimensional systems of closely spaced metallic nanoparticles exhibiting LSPR in their optical response. The thesis work has experimentally explored the extent to which macroscopic, externally tunable fabrication parameters can influence the morphology and the arrangement of the systems' components at the nanometric scale effectively enabling the morphology-mediated, a priori control of the plasmonic response. The control of optical characteristics such as width and energy of the LSPR and the degree of birefringence of the two-dimensional plasmonic medium is experimentally demonstrated, and its underlying mecha-nisms explained by means of ad hoc models.

The thesis work has addressed what likely represents the simplest and most commonly available plasmonic materials, gold, proving the general feasibility of the method and disclosing the potential of the self-organization approach. Based on this work, the use of even more sophisticated fabrication schemes opens new exciting possibilities for these self-organized systems. Recent research addressed the use of materials other than gold as the plasmonic counterpart, like the pure metals Ag and especially Aluminium, that allowed taking the plasmonic functionality into the deep ultraviolet region. Alloy nanoparticles with a freely variable composition add a further degree of freedom to the tunability of the systems' response, allowing to flexibly manipulate the plasmonic response over a very large frequency range. Substrates with broadband tunable LSPR can be exploited as flexible supports for plasmon-mediated excitations of deposited molecules or quantum dots, where the capability of addressing, within a single system, in- and off-resonance behaviors has the potential to highlight the plasmon role in mediating the optical excitation.

Genoa, March 2012   Dr. Francesco Bisio

Acknowledgments

First of all, I would like to thank Dr. Francesco Bisio and Prof. Lorenzo Mattera for making this work possible and for their precious guidance. A big thank you also to Prof. Maurizio Canepa for his useful advice and discussions. Thanks to Dr. Riccardo Moroni for his assistance in the experiments, and to Ennio Vigo for the technical support during the realization of the preparation chamber. Grateful thanks also to Dr. Mirko Prato for his precious help on ellipsometry.

I gratefully acknowledge the group of biophysics, in particular Prof. Ornella Cavalleri, for allowing me to use the AFM and DLS instrumentations and access the chemistry laboratory. A particular thank you also to Dr. Amanda Penco for her kind assistance in the AFM measurements.

I deeply thank the group of Maria Del Puerto Morales at CSIC/ICCM of Madrid for the collaboration in the synthesis of the magnetic nanoparticles.

Finally, I would like to thank Laura Caprile, Michael Caminale, Dr. Elena Gatta, and Chiara Toccafondi for making my thesis a pleasant experience.

Contents

Chapter 1
Introduction

In a world in which the information and communication technology has gradually assumed a pivotal role for scientific, technological and social purposes, there is a constantly growing demand for smaller and faster devices able of manipulating information, typically in the form of electronic, magnetic or optical signals. The request of smaller and faster devices has inevitably led to a constant process of miniaturization of their elementary components, thereby bringing along a huge load of scientific and technological challenges that researchers and engineers have to face. Nowadays, for example, state-of-the-art technology requires excellent control of structures having typical lateral dimension of the order of approximately ten nanometers, not too far off molecular dimensions.

While conventional lithographic methods [1–6] always represent the standard choice for the fabrication of technological nanostructures, the challenges associated with the constant size reduction have promoted intense research aimed at exploiting alternative structures for the fabrication of nanosized systems [7–20].

Among the various strategies, in the past years we have witnessed a growing interest to the world of colloid and cluster science, leading to the consolidation of nanoparticles (NPs) as convenient structural elements for the construction of functional interfaces [10, 14, 21–28]. In fact, nanoparticles can be fashioned from many different materials, presenting a wide diversity of electronic, optical, catalytic and magnetic properties, which in many cases originate from the reduced dimensionality of the systems and thus are not found in the bulk counterparts [29–34].

While the properties of individual NPs can be exploited in a variety of ways [8, 21, 23, 26, 35–47], other interesting functionalities arise when the NPs are assembled in such a way that each one "feels" the presence of neighbouring particles via some mutual interaction. Typical examples of systems exhibiting such collective functionality are assemblies of magnetic particles [14, 27, 28, 34, 48–50], interacting with one another through their magnetic dipolar field or, most relevant for this thesis, assemblies of metallic particles characterized by peculiar optical functionalities [24, 25, 38, 51–60].

L. Anghinolfi, *Self-Organized Arrays of Gold Nanoparticles*, Springer Theses, DOI: 10.1007/978-3-642-30496-5_1, © Springer-Verlag Berlin Heidelberg 2012

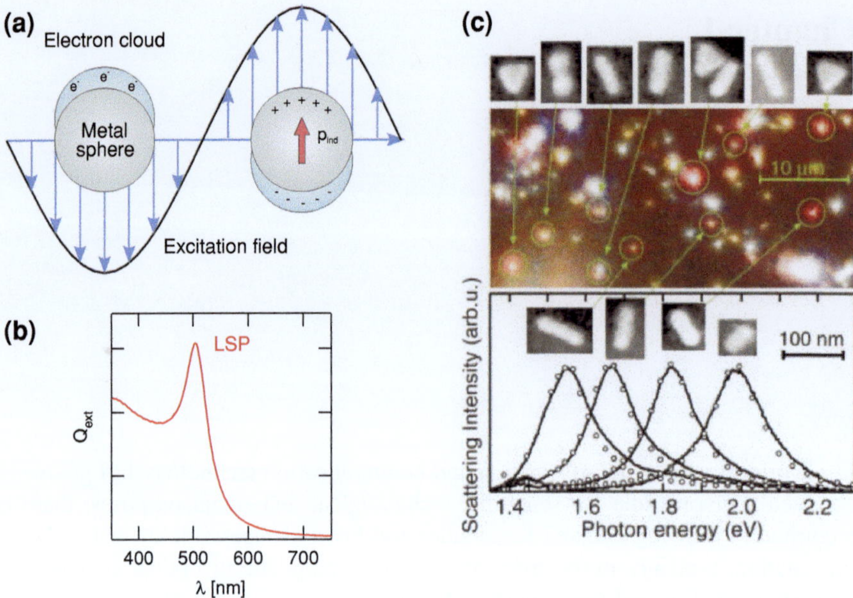

Fig. 1.1 Panel **a** Schematics of a metal nanoparticle illuminated by an electromagnetic wave; the electric field displaces the electronic cloud inducing an electric dipole inside the particle. **b** Calculated extinction coefficient for a gold nanoparticle; the peak at ≈520 nm corresponds to the particle's resonant scattering and absorption of light, i.e. to the excitation of the LSP. **c** Dark field microscopy image (*top*) and scattering spectra (*bottom*) of Au nanoparticles with different shapes (reprinted with permission from Ref. [61]. Copyright 2003, American Institute of Physics)

Under the influence of the electromagnetic (EM) field, metallic nanoparticles can in fact exhibit strong resonant absorptions (Fig. 1.1a, b), absent in bulk counterparts, referred to as localized surface plasmon resonances (LSPRs) [54, 62–66]. A detailed description of the interaction between EM waves and metal nanostructures can be found in several standard text books, like Refs. [67–69]. The main features of LSPs (i.e. the characteristic frequency and the width and the intensity of the resonance) are strongly sensitive to both intrinsic geometrical factors, like the NP shape and size (Fig. 1.1c), and external variables, like the NP dielectric environment [54, 61, 63, 70, 71].

The near proximity of the NPs to another metallic material, in the form of a surface or of other neighbouring NPs, causes dramatic variations of the LSPR characteristics driven by the near-field EM coupling between the NP and its metallic surroundings [63, 75–91] (Fig. 1.2a). Beside modifying the LSPR, EM coupling leads to interesting properties in terms of localization, enhancement and guiding of the EM field on subwavelength scales [59, 92, 93]. For example, 1-dimensional (1D) chains of NPs are extremely appealing for their capability of confining the EM energy below the diffraction limit and for low-loss, high-speed EM signal transmission (Fig. 1.2b, c), a property that can be exploited in hybrid plasmonic/electronic devices [58, 94, 95].

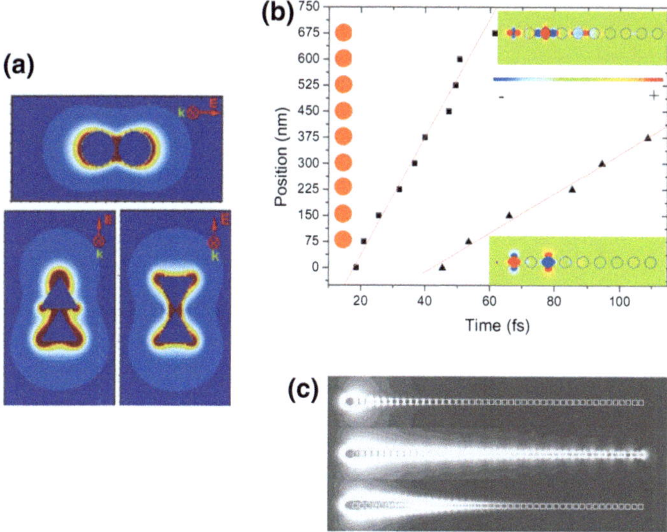

Fig. 1.2 Panel **a** (reprinted with permission from Ref. [72]. Copyright 2004, American Institute of Physics): calculated electric near-field profiles external to Ag nanoprism pairs with different shapes (the full scale is 4 orders of magnitudes with respect to the exciting field). Panel **b** (reprinted with permission from Ref. [73]. Copyright 2003 by the American Physical Society): transmission of a electromagnetic pulse centered at the LSP frequency through a Ag nanoparticle plasmon waveguide; the graph shows the FDTD-calculated pulse-peak position versus time for longitudinal (*black squares*) and transverse (*black triangles*) polarization; snapshots of the corresponding in-plane electric fields are shown in the insets. **c** (reprinted with permission from Ref. [74]. Copyright 2006 IEEE): electric field distribution in a Ag nanoparticle plasmon waveguide excited to the left end; only when the waveguide is excited at the LSP resonance (*central row*), the energy is efficiently propagated along the array

Extensive reviews on the topic of plasmonic waveguides can be found in Refs. [92, 93].

In general, the collective properties of ordered ensembles of NPs stem from a superposition of single-NP, intrinsic geometrical factors (shape, size, orientation with respect to the exciting field), and ensemble properties (interparticle spacing, symmetry etc.) [82–91]. Changing either of the two classes of factors leads to a corresponding modification of the collective response of the systems, thereby offering the intriguing possibility of tailoring the functionality of NP ensembles according to the specific scientific/technological requirements [54, 96], like field enhancement for nonlinear spectroscopies [97], EM signal transmission [59], sensing [55] etc.

Clearly, in order to make the most of the collective properties of ensembles of plasmonic nanoparticles, a strategy for their assembly must be found, which allows to flexibly modify structural parameters of the system like mean particle-particle spacing, or the ensemble geometry. Focusing on bottom-up approaches, based on self-assembly or self-organized methods, several fabrication strategies can be found. In the case of particle deposition onto a 2-dimensional support, in the purest bottom-up approach, the particles deposited on a flat surface will tend to assume a close-packed geometry [14, 98–101]. This leaves a relatively restricted number of ways to modify their arrangement: for example the particle-particle spacing can be modified by varying the length of the surfactant molecules [14, 23, 101], or the array symmetry can be varied changing the shape of the particles [102–106]. Another possibility is instead to employ nanopatterned templates [107, 108] as supports, so that the nanoparticles arrangement can be guided in a controllable manner. Nanopatterned templates can be realized, for example, by creating a chemical contrast pattern between regions with different affinity to the nanoparticles [109–112], by printing processes [7, 9, 22, 113], or by inducing periodical modifications of the surface morphology [114–122]; the combined application of nanolithography, for the pattering of the substrates, and self-assembling, for the deposition of nanoparticles, is also often considered [123–126].

In this thesis, we report the fabrication of self-organized 2-dimensional (2D) arrays of Au nanoparticles with tunable particle shape and mutual arrangement, realized employing nanostructured LiF(110) substrates as patterned templates. Insulating ionic crystals, like NaCl, LiF, CaF_2 or MgO, have already been proposed since more than a decade as suitable templates for nanopatterned assembly [115, 127–131]. In fact, faceting of the (110)- or (111)-like surfaces into ridge-and-valley or pyramidal structures can be spontaneously induced by means of homoepitaxial deposition or mild annealing. The same pattern can then be easily replicated by depositing a thin layer of the material of interest, allowing, for example, to realize arrays of magnetic nanowires and nanodots [115, 132, 133], or of gold nanowires for second harmonic generation [134]. Here we apply this technique to the investigation of the collective optical response of a 2D arrays of gold nanodots.

Our 2D arrays of gold nanoparticles were fabricated by deposition of Au onto the self-organized nanoscale ridge-and-valley morphological patterns that spontaneously form upon homoepitaxial deposition at the LiF(110) surface. Grazing-incidence evaporation of Au onto the LiF nanopatterns leads to the formation of Au nanowires, that evolve towards regular arrays of disconnected NPs, aligned with the LiF ridges, following a temperature-induced dewetting. Depending on the substrate fabrication and the Au deposition parameters, the NP shape (coherently-aligned ellipsoids with tunable aspect ratio and size) and the array characteristics (interparticle spacing, array symmetry) could be independently controlled, allowing to correspondingly tune the system's plasmonic response.

In this work, we demonstrate the potential of the method by focusing on two particularly relevant cases, namely circular NPs arranged on a rectangular lattice, and coherently-aligned elongated ellipsoids laid on a square mesh. Each system is endowed with one single specific symmetry-breaking characteristic, the array symmetry in the former case and the NP shape in the latter, and both exhibit in-plane optical birefringence. In the first case, the birefringence has its roots exclusively in the anisotropic electromagnetic coupling between the NP, arising from the uniaxial symmetry of the rectangular lattice. In the second case, the intrinsic anisotropic response of each NP to the exciting field provides a double contribution to the optical birefringence, firstly via an intrinsic anisotropic polarizability of each Au NP and secondly via the consequently anisotropic EM dipole radiated field.

The experimental findings are discussed within a frame of a simple yet comprehensive effective-medium model, that quantitatively accounts for NP shape, EM coupling and substrate effects, reproducing the experimental observations and allowing to rationalize the intrinsic and collective effects that concur in determining the system's optical response. We show that arrays endowed with elongated ellipsoidal NPs allow a greater flexibility in the engineering of the degree of birefringence in the collective plasmonic response. The reported methods and analysis thus provide a simple route for the cheap fabrication of large-area plasmonic systems with tailored SPR characteristics, exploitable as tunable supports for SPR-enhanced optical spectroscopy [97] or SPR-based sensing [55].

This thesis is structured as follows:

Chapter 2 The general aspects of the interaction between light and matter will be reviewed, focusing the attention on the analytical description of heterogeneous materials and on the optical response of metallic nanoparticles.

Chapter 3 A detailed description of the experimental apparatus, and a brief introduction to the experimental methods employed for the characterization of the samples, will be presented.

Chapter 4 The procedure for the fabrication of the 2D arrays of gold nanoparticles will be presented, reporting the morphological characterization of the nanopatterned LiF(110) substrates and of the arrays as a function of the growth parameters.

Chapter 5 The optical characterizations of the samples, by means of spectroscopic ellipsometry, reflectivity and transmissivity, are reported at each step of the fabrication procedure, with particular emphasis on the characteristics of the collective plasmonic resonances in the arrays.

Chapter 6 The optical measurements are compared to model calculations, in order to associate the optical features to the morphology of the samples. In particular, we develop a theoretical framework to describe the optical response of the gold nanoparticles arrays, and then apply it to two selected samples in order to separate the contributions to the plasmonic response.

Chapter 7 The 2D arrays of gold nanoparticles are employed as templates for the guided deposition of magnetite nanoparticles, for the realization of an optically active device. Optical and morphological characterizations after the deposition from a colloidal suspension are described.

References

1. Stephen Y. Chou, Peter R. Krauss, and Preston J. Renstrom. Imprint of sub-25 nm vias and trenches in polymers. *App. Phys. Lett.*, 67:3114, 1995.
2. Stephen Y. Chou, Peter R. Krauss, and Preston J. Renstrom. Imprint lithography with 25-nanometer resolution. *Science*, 272:85, 1996.
3. Matthew Colburn, Stephen C. Johnson, Michael D. Stewart, S. Damle, Todd C. Bailey, Bernard Choi, M. Wedlake, Timothy B. Michaelson, S. V. Sreenivasan, John G. Ekerdt, and C. Grant

Willson. Step and flash imprint lithography: a new approach to high-resolution patterning. volume 3676, page 379. SPIE, 1999.

4. Saleem H. Zaidi and S. R. J. Brueck. Multiple-exposure interferometric lithography. *J. Vac. Sci. Tech. B*, 11:658, 1993.

5. Kenji Gamo. Nanofabrication by FIB. *Microelectr. Eng.*, 32:159, 1996.

6. John C. Hulteen and Richard P. Van Duyne. Nanosphere lithography: A materials general fabrication process for periodic particle array surfaces. *J. Vac. Sci. Tech. A*, 13:1553, 1995.

7. Richard D. Piner, Jin Zhu, Feng Xu, Seunghun Hong, and Chad A. Mirkin. "Dip-Pen" nano-lithography. *Science*, 283:661, 1999.

8. Shuguang Zhang. Fabrication of novel biomaterials through molecular self-assembly. *Nat. Biotech.*, 21:1171, 2003.

9. Ki-Bum Lee, So-Jung Park, Chad A. Mirkin, Jennifer C. Smith, and Milan Mrksich. Protein nanoarrays generated by dip-pen nanolithography. *Science*, 295:1702, 2002.

10. J. Christopher Love, Lara A. Estroff, Jennah K. Kriebel, Ralph G. Nuzzo, and George M. Whitesides. Self-assembled monolayers of thiolates on metals as a form of nanotechnology. *Chem. Rev.*, 105:1103, 2005.

11. Michael J Fasolka and Anne M Mayes. Block copolymer thin films: Physics and applications. *Ann. Rev. Mat. Res.*, 31:323, 2001.

12. M. P. Pileni, Y. Lalatonne, D. Ingert, I. Lisiecki, and A. Courty. Self assemblies of nanocrystals: preparation, collective properties and uses. *Faraday Discuss.*, 125:251, 2004.

13. C. B. Murray, C. R. Kagan, and M. G. Bawendi. Synthesis and characterization of monodisperse nanocrystals and close-packed nanocrystal assemblies. *Annual Review of Materials Science*, 30:545, 2000.

14. Shouheng Sun, C. B. Murray, Dieter Weller, Liesl Folks, and Andreas Moser. Monodisperse FePt nanoparticles and ferromagnetic FePt nanocrystal superlattices. *Science*, 287(5460):1989, 2000.

15. Eunhee Jeoung, Trent H. Galow, Joerg Schotter, Mustafa Bal, Andrei Ursache, Mark T. Tuominen, Christopher M. Stafford, Thomas P. Russell, and Vincent M. Rotello. Fabrication and characterization of nanoelectrode arrays formed via block copolymer self-assembly. *Langmuir*, 17:6396, 2001.

16. Zhaoxiang Deng and Chengde Mao. Molecular lithography with DNA nanostructures. *Ang. Chem. Int. Ed.*, 43:4068, 2004.

17. Mato Knez, Alexander M. Bittner, Fabian Boes, Christina Wege, Holger Jeske, E. Mai, and Klaus Kern. Biotemplate synthesis of 3-nm nickel and cobalt nanowires. *Nano Letters*, 3:1079, 2003.

18. Zhaoxiang Deng and Chengde Mao. DNA-templated fabrication of 1D parallel and 2D crossed metallic nanowire arrays. *Nano Letters*, 3:1545, 2003.

19. Hao Yan, Sung Ha Park, Gleb Finkelstein, John H. Reif, and Thomas H. LaBean. DNA-templated self-assembly of protein arrays and highly conductive nanowires. *Science*, 301:1882, 2003.

20. Chad A. Mirkin, Robert L. Letsinger, Robert C. Mucic, and James J. Storhoff. A DNA-based method for rationally assembling nanoparticles into macroscopic materials. *Nature*, 382:607, 1996.

21. Jwa-Min Nam, C. Shad Thaxton, and Chad A. Mirkin. Nanoparticle-based bio-bar codes for the ultrasensitive detection of proteins. *Science*, 301:1884, 2003.

22. Ki-Bum Lee, Jung-Hyurk Lim, and Chad A. Mirkin. Protein nanostructures formed via direct-write dip-pen nanolithography. *J. Am. Chem. Soc.*, 125:5588, 2003.

23. Marie-Christine Daniel and Didier Astruc. Gold nanoparticles: Assembly, supramolecular chemistry, quantum-size-related properties, and applications toward biology, catalysis, and nanotechnology. *Chem. Rev.*, 104:293, 2004.

24. L. A. Sweatlock, S. A. Maier, H. A. Atwater, J. J. Penninkhof, and A. Polman. Highly confined electromagnetic fields in arrays of strongly coupled Ag nanoparticles. *Phys. Rev. B*, 71:235408, 2005.

25. Amanda J. Haes, W. Paige Hall, Lei Chang, William L. Klein, and Richard P. Van Duyne. A localized surface plasmon resonance biosensor: First steps toward an assay for Alzheimer's disease. *Nano Letters*, 4:1029, 2004.

26. Anke S. Wrz, Ken Judai, Stphane Abbet, and Ulrich Heiz. Cluster size-dependent mechanisms of the CO + NO reaction on small Pd_n (n = 30) clusters on oxide surfaces. *J. Am. Chem. Soc.*, 125:7964, 2003.

27. B. Warne, O.I. Kasyutich, E.L. Mayes, J.A.L. Wiggins, and K.K.W. Wong. Self assembled nanoparticulate Co:Pt for data storage applications. *IEEE Trans. Mag.*, 36:3009, 2000.

28. S. Sun. Recent advances in chemical synthesis, self-assembly, and applications of FePt nanoparticles. *Adv. Mat.*, 18:393, 2006.

29. Tito Trindade, Paul O'Brien, and Nigel L. Pickett. Nanocrystalline semiconductors: Synthesis, properties, and perspectives. *Chem. Mat.*, 13:3843, 2001.

30. Marcos M. Alvarez, Joseph T. Khoury, T. Gregory Schaaff, Marat N. Shafigullin, Igor Vezmar, and Robert L. Whetten. Optical absorption spectra of nanocrystal gold molecules. *J. Phys. Chem. B*, 101:3706, 1997.

31. R. H. Kodama, Salah A. Makhlouf, and A. E. Berkowitz. Finite size effects in antiferromagnetic NiO nanoparticles. *Phys. Rev. Lett.*, 79:1393, 1997.

32. M. Respaud, J. M. Broto, H. Rakoto, A. R. Fert, L. Thomas, B. Barbara, M. Verelst, E. Snoeck, P. Lecante, A. Mosset, J. Osuna, T. Ould Ely, C. Amiens, and B. Chaudret. Surface effects on the magnetic properties of ultrafine cobalt particles. *Phys. Rev. B*, 57:2925, 1998.

33. G. F. Goya, T. S. Berquó, F. C. Fonseca, and M. P. Morales. Static and dynamic magnetic properties of spherical magnetite nanoparticles. *J. App. Phys.*, 94:3520, 2003.

34. J. Bansmann, S.H. Baker, C. Binns, J.A. Blackman, J.-P. Bucher, J. Dorantes-Dvila, V. Dupuis, L. Favre, D. Kechrakos, A. Kleibert, K.-H. Meiwes-Broer, G.M. Pastor, A. Perez, O. Toulemonde, K.N. Trohidou, J. Tuaillon, and Y. Xie. Magnetic and structural properties of isolated and assembled clusters. *Surf. Sci. Rep.*, 56:189, 2005.

35. Nathaniel L. Rosi and Chad A. Mirkin. Nanostructures in biodiagnostics. *Chem. Rev.*, 105:1547, 2005.

36. Prashant V. Kamat. Photophysical, photochemical and photocatalytic aspects of metal nanoparticles. *J. Phys. Chem. B*, 106:7729, 2002.

37. Mauro Ferrari. Cancer nanotechnology: opportunities and challenges. *Nature Reviews. Cancer*, 5:161, 2005.

38. Adam D. McFarland and Richard P. Van Duyne. Single silver nanoparticles as real-time optical sensors with zeptomole sensitivity. *Nano Letters*, 3:1057, 2003.

39. YunWei Charles Cao, Rongchao Jin, and Chad A. Mirkin. Nanoparticles with Raman spectroscopic fingerprints for DNA and RNA detection. *Science*, 297:1536, 2002.

40. Jayanth Panyam and Vinod Labhasetwar. Biodegradable nanoparticles for drug and gene delivery to cells and tissue. *Adv. Drug Deliv. Rev.*, 55:329, 2003.

41. Krishnendu Roy, Hai-Quan Mao, Shau-Ku Huang, and Kam W. Leong. Oral gene delivery with chitosan-DNA nanoparticles generates immunologic protection in a murine model of peanut allergy. *Nature Medicine*, 5:387, 1999.

42. Christof M. Niemeyer. Nanoparticles, proteins, and nucleic acids: Biotechnology meets materials science. *Ang. Chem. Int. Ed.*, 40:4128, 2001.

43. Ajay Kumar Gupta and Mona Gupta. Synthesis and surface engineering of iron oxide nanoparticles for biomedical applications. *Biomaterials*, 26:3995, 2005.

44. Q A Pankhurst, J Connolly, S K Jones, and J Dobson. Applications of magnetic nanoparticles in biomedicine. *J. Phys. D*, 36:R167, 2003.

45. Xiaohua Huang, Ivan H. El-Sayed, Wei Qian, and Mostafa A. El-Sayed. Cancer cell imaging and photothermal therapy in the near-infrared region by using gold nanorods. *J. Am. Chem. Soc.*, 128:2115, 2006.

46. Stephane Mornet, Sebastien Vasseur, Fabien Grasset, and Etienne Duguet. Magnetic nanoparticle design for medical diagnosis and therapy. *J. Mater. Chem.*, 14:2161, 2004.

47. Andreas Jordan, Regina Scholz, Peter Wust, Horst Fähling, and Roland Felix. Magnetic fluid hyperthermia (MFH): Cancer treatment with AC magnetic field induced excitation of biocompatible superparamagnetic nanoparticles. *J. Mag. Mag. Mat.*, 201:413, 1999.

48. B D Terris and T Thomson. Nanofabricated and self-assembled magnetic structures as data storage media. *J. Phys. D*, 38:R199, 2005.

49. Xiaowei Teng and Hong Yang. Synthesis of face-centered tetragonal FePt nanoparticles and granular films from Pt@Fe$_2$O$_3$ core-shell nanoparticles. *J. Am. Chem. Soc.*, 125:14559, 2003.

50. G. A. Held, Hao Zeng, and Shouheng Sun. Magnetics of ultrathin FePt nanoparticle films. *J. App. Phys.*, 95:1481, 2004.

51. Ekmel Ozbay. Plasmonics: Merging photonics and electronics at nanoscale dimensions. *Science*, 311:189, 2006.

52. Liang-shi Li, Joost Walda, Liberato Manna, and A. Paul Alivisatos. Semiconductor nanorod liquid crystals. *Nano Letters*, 2:557, 2002.

53. Marisa Mäder, Thomas Höche, Jürgen W. Gerlach, Susanne Perlt, Jens Dorfmüller, Michael Saliba, Ralf Vogelgesang, Klaus Kern, and Bernd Rauschenbach. Plasmonic activity of large-area gold nanodot arrays on arbitrary substrates. *Nano Letters*, 10:47, 2010.

54. E. Hutter and J. H. Fendler. Exploitation of localized surface plasmon resonance. *Adv. Mater.*, 16:1685, 2004.

55. J. N. Anker, W. P. Hall, O. Lyandres, N. C. Shah, J. Zhao, and R. P. Van Duyne. Biosensing with plasmonic nanosensors. *Nature. Mater.*, 7:442, 2008.

56. Amanda J. Haes, Shengli Zou, George C. Schatz, and Richard P. Van Duyne. A nanoscale optical biosensor: The long range distance dependence of the localized surface plasmon resonance of noble metal nanoparticles. *J. Phys. Chem. B*, 108:109, 2004.

57. Christopher J. Orendorff, Anand Gole, Tapan K. Sau, and Catherine J. Murphy. Surface-enhanced Raman spectroscopy of self-assembled monolayers: Sandwich architecture and nanoparticle shape dependence. *Anal. Chem.*, 77:3261, 2005.

58. Stefan A. Maier, Pieter G. Kik, and Harry A. Atwater. Observation of coupled plasmon-polariton modes in Au nanoparticle chain waveguides of different lengths: Estimation of waveguide loss. *App. Phys. Lett.*, 81:1714, 2002.

59. S. A. Maier, P. G. Kik, H. A. Atwater, S. Meltzer, E. Harel, B. E. Koel, and A. A. G. Requicha. Local detection of electromagnetic energy transport below the diffraction limit in metal nanoparticle plasmon waveguides. *Nature Mater.*, 2:229, 2003.

60. Stefan A. Maier, Mark L. Brongersma, Pieter G. Kik, and Harry A. Atwater. Observation of near-field coupling in metal nanoparticle chains using far-field polarization spectroscopy. *Phys. Rev. B*, 65:193408, 2002.

61. Hitoshi Kuwata, Hiroharu Tamaru, Kunio Esumi, and Kenjiro Miyano. Resonant light scattering from metal nanoparticles: Practical analysis beyond rayleigh approximation. *Appl. Phys. Lett.*, 83:4625, 2003.

62. U. Kreibig. Optical absorption of small metallic particles. *Surf. Sci.*, 156:678, 1985.

63. Cecilia Noguez. Surface plasmons on metal nanoparticles: The influence of shape and physical environment. *J. Phys. Chem. C*, 111:3806, 2007.

64. Poelsema B. Stefan Kooij E. Shape and size effects in the optical properties of metallic nanorods. *Phys. Chem. Chem. Phys.*, 8:3349, 2006.

65. S. Link, M. B. Mohamed, and M. A. El-Sayed. Simulation of the optical absorption spectra of gold nanorods as a function of their aspect ratio and the effect of the medium dielectric constant. *J. Phys. Chem. B*, 103:3073, 1999.

66. Stephan Link and Mostafa A. El-Sayed. Size and temperature dependence of the plasmon absorption of colloidal gold nanoparticles. *J. Phys. Chem. B*, 103:4212, 1999.

67. Bohren C.F.; Huffman D.R. *Absorption and scattering of light by small particles*. Wiley, 1998.

68. U. Kreibig and M. Vollmer. *Optical Properties of Metal Clusters*. Springer, 1995.

69. M. Quinten. *Optical Properties of Nanoparticle Systems*. Wiley-VCH, Berlin, 2011.

70. K. Lance Kelly, Eduardo Coronado, Lin Lin Zhao, and George C. Schatz. The optical properties of metal nanoparticles: The influence of size, shape, and dielectric environment. *J. Phys. Chem. B*, 107:668, 2003.

71. Molly M. Miller and Anne A. Lazarides. Sensitivity of metal nanoparticle surface plasmon resonance to the dielectric environment. *J. Phys. Chem. B*, 109:21556, 2005.

72. Encai Hao and George C. Schatz. Electromagnetic fields around silver nanoparticles and dimers. *J. Chem. Phys.*, 120:357, 2004.

73. Stefan A. Maier, Pieter G. Kik, and Harry A. Atwater. Optical pulse propagation in metal nanoparticle chain waveguides. *Phys. Rev. B*, 67:205402, 2003.

74. S. A. Maier. Plasmonics: Metal nanostructures for subwavelength photonic devices. *Selected Topics in Quantum Electronics, IEEE Journal of*, 12:1214, 2006.

75. K.-H. Su, Q.-H. Wei, X. Zhang, J. J. Mock, D. R. Smith, and S. Schultz. Interparticle coupling effects on plasmon resonances of nanogold particles. *Nano Letters*, 3:1087, 2003.

76. W. Rechberger, A. Hohenau, A. Leitner, J. R. Krenn, B. Lamprecht, and F. R. Aussenegg. Optical properties of two interacting gold nanoparticles. *Opt. Comm.*, 220:137, 2003.

77. K. George Thomas, Said Barazzouk, Binil Itty Ipe, S. T. Shibu Joseph, and Prashant V. Kamat. Uniaxial plasmon coupling through longitudinal self-assembly of gold nanorods. *J. Phys. Chem. B*, 108:13066, 2004.

78. Prashant K. Jain, Wenyu Huang, and Mostafa A. El-Sayed. On the universal scaling behavior of the distance decay of plasmon coupling in metal nanoparticle pairs: A plasmon ruler equation. *Nano Letters*, 7:2080, 2007.

79. Christopher Tabor, Desiree Van Haute, and Mostafa A. El-Sayed. Effect of orientation on plasmonic coupling between gold nanorods. *ACS Nano*, 3:3670, 2009.

80. Christopher Tabor, Raghunath Murali, Mahmoud Mahmoud, and Mostafa A. El-Sayed. On the use of plasmonic nanoparticle pairs as a plasmon ruler: The dependence of the near-field dipole plasmon coupling on nanoparticle size and shape. *J. Phys. Chem. A*, 113:1946, 2009.

81. Prashant K. Jain and Mostafa A. El-Sayed. Plasmonic coupling in noble metal nanostructures. *Chem. Phys. Lett.*, 487:153, 2010.

82. A. Taleb, V. Russier, A. Courty, and M. P. Pileni. Collective optical properties of silver nanoparticles organized in two-dimensional superlattices. *Phys. Rev. B*, 59:13350, 1999.

83. L. L. Zhao, K. L. Kelly, and G. C. Schatz. The extinction spectra of silver nanoparticle arrays: Influence of array structure on plasmon resonance wavelength and width. *J. Phys. Chem. B*, 107:7343, 2003.

84. Christy L. Haynes, Adam D. McFarland, LinLin Zhao, Richard P. Van Duyne, George C. Schatz, Linda Gunnarsson, Juris Prikulis, Bengt Kasemo, and Mikael Käll. Nanoparticle optics: The importance of radiative dipole coupling in two-dimensional nanoparticle arrays. *J. Phys. Chem. B*, 107:7337, 2003.

85. Erin M. Hicks, Shengli Zou, George C. Schatz, Kenneth G. Spears, Richard P. Van Duyne, Linda Gunnarsson, Tomas Rindzevicius, Bengt Kasemo, and Mikael Käll. Controlling plasmon line shapes through diffractive coupling in linear arrays of cylindrical nanoparticles fabricated by electron beam lithography. *Nano Lett.*, 5:1065, 2005.

86. Luis M. Liz-Marzán. Tailoring surface plasmons through the morphology and assembly of metal nanoparticles. *Langmuir*, 22:32, 2006.

87. S. K. Ghosh and T. Pal. Interparticle coupling effect on the surface plasmon resonance of gold nanoparticles: From theory to applications. *Chem. Rev.*, 107:4797, 2007.

88. T. W. H. Oates, A. Keller, S. Faksko, and A. Mücklich. Aligned silver nanoparticles on rippled silicon templates exhibiting anisotropic plasmon absorption. *Plasmonics*, 2:47, 2007.

89. Stéphanie Vial, Isabel Pastoriza-Santos, Jorge Pérez-Juste, and Luis M. Liz-Marzán. Plasmon coupling in layer-by-layer assembled gold nanorod films. *Langmuir*, 23:4606, 2007.

90. Fredrik Westerlund and Thomas Björnholm. Directed assembly of gold nanoparticles. *Curr. Opin. Coll. Interf.*, 14:126, 2009.

91. I. Romero and F. J. García de Abajo. Anisotropy and particle-size effects in nanostructured plasmonic metamaterials. *Opt. Expr.*, 17:22010, 2009.

92. S. A. Maier. *Plasmonics: Fundamentals and Applications*. Springer, 2007.

93. M. L. Brongersma and P. G. Kik, editors. *Surface Plasmon Nanophotonics*. Springer, 2007.

94. M. Quinten, A. Leitner, J. R. Krenn, and F. R. Aussenegg. Electromagnetic energy transport via linear chains of silver nanoparticles. *Opt. Lett.*, 23:1331, 1998.

95. Mark L. Brongersma, John W. Hartman, and Harry A. Atwater. Electromagnetic energy transfer and switching in nanoparticle chain arrays below the diffraction limit. *Phys. Rev. B*, 62:R16356, 2000.

96. Andrew N. Shipway, Eugenii Katz, and Itamar Willner. Nanoparticle arrays on surfaces for electronic, optical, and sensor applications. *ChemPhysChem*, 1:18, 2000.

97. H. Ko, S. Singamaneni, and V. V. Tsukruk. Nanostructured surfaces and assemblies as SERS media. *Small*, 4:1576, 2008.

98. C. Petit,, and M. P. Pileni. Cobalt nanosized particles organized in a 2D superlattice: Synthesis, characterization, and magnetic properties. *J. Phys. Chem. B*, 103:1805, 1999.

99. M. P. Pileni. Nanocrystal self-assemblies: Fabrication and collective properties. *J. Phys. Chem. B*, 105:3358, 2001.

100. N. Shukla, Erik B. Svedberg, and J. Ell. Long range ordering of self-assembled monolayers of FePt nanoparticles on modified substrates. *Surf. Coat. Tech.*, 201:3810, 2006.

101. Takami Shimizu, Toshiharu Teranishi, Satoshi Hasegawa, and Mikio Miyake. Size evolution of alkanethiol-protected gold nanoparticles by heat treatment in the solid state. *J. Phys. Chem. B*, 107:2719, 2003.

102. David D. Evanoff and George Chumanov. Synthesis and optical properties of silver nanoparticles and arrays. *ChemPhysChem*, 6:1221, 2005.

103. Catherine J. Murphy, Tapan K. Sau, Anand M. Gole, Christopher J. Orendorff, Jinxin Gao, Linfeng Gou, Simona E. Hunyadi, and Tan Li. Anisotropic metal nanoparticles: Synthesis, assembly, and optical applications. *J. Phys. Chem. B*, 109:13857, 2005.

104. Franklin Kim, Serena Kwan, Jennifer Akana, and Peidong Yang. Langmuir-Blodgett nanorod assembly. *Journal of the American Chemical Society*, 123:4360, 2001.

105. Sihai Chen, Zhiyong Fan, and David L. Carroll. Silver nanodisks: Synthesis, characterization, and self-assembly. *J. Phys. Chem. B*, 106:10777, 2002.

106. A. N. Lagarkov and A. K. Sarychev. Electromagnetic properties of composites containing elongated conducting inclusions. *Phys. Rev. B*, 53:6318, 1996.

107. Rachel K. Smith, Penelope A. Lewis, and Paul S. Weiss. Patterning self-assembled monolayers. *Progr. Surf. Sci.*, 75:1, 2004.

108. M. Geissler and Y. Xia. Patterning: Principles and some new developments. *Adv. Mat.*, 16:1249, 2004.

109. Xing Ling, Xin Zhu, Jin Zhang, Tao Zhu, Manhong Liu, Lianming Tong, and Zhongfan Liu. Reproducible patterning of single Au nanoparticles on silicon substrates by scanning probe oxidation and self-assembly. *J. Phys. Chem. B*, 109:2657, 2005.

110. Shantang Liu, Rivka Maoz, and Jacob Sagiv. Planned nanostructures of colloidal gold via self-assembly on hierarchically assembled organic bilayer template patterns with in-situ generated terminal amino functionality. *Nano Letters*, 4:845, 2004.

111. Nisha Shukla, Erik B. Svedberg, and John Ell. FePt nanoparticle adsorption on a chemically patterned silicon-gold substrate. *Surf. Coat. Tech.*, 201:1256, 2006.

112. Joanna Aizenberg, Andrew J. Black, and George M. Whitesides. Controlling local disorder in self-assembled monolayers by patterning the topography of their metallic supports. *Nature*, 394:868, 1998.

113. Tobias Kraus, Laurent Malaquin, Heinz Schmid, Walter Riess, Nicholas D. Spencer, Nicholas D. Spencer, Nicholas D. Spencer, and Heiko Wolf. Nanoparticle printing with single-particle resolution. *Nat. Nano.*, 2:570, 2007.

114. Toshiharu Teranishi, Akira Sugawara, Takami Shimizu, and Mikio Miyake. Planar array of 1D gold nanoparticles on ridge-and-valley structured carbon. *J. Am. Chem. Soc.*, 124:4210, 2002.

115. Akira Sugawara, G. G. Hembree, and M. R. Scheinfein. Self-organized mesoscopic magnetic structures. *J. App. Phys.*, 82:5662, 1997.

116. Akira Sugawara and K. Mae. Faceting of homoepitaxial MgO(110) layers prepared by electron beam evaporation. *Surf. Sci.*, 558:211, 2004.

117. Michael P. Zach, Kwok H. Ng, and Reginald M. Penner. Molybdenum nanowires by electrodeposition. *Science*, 290:2120, 2000.

118. J. Dekoster, B. Degroote, H. Pattyn, G. Langouche, A. Vantomme, and S. Degroote. Step decoration during deposition of Co on Ag(001) by ultralow energy ion beams. *App. Phys. Lett.*, 75:938, 1999.

119. P. Gambardella, M. Blanc, H. Brune, K. Kuhnke, and K. Kern. One-dimensional metal chains on Pt vicinal surfaces. *Phys. Rev. B*, 61:2254, 2000.

120. D. Y. Petrovykh, F. J. Himpsel, and T. Jung. Width distribution of nanowires grown by step decoration. *Surf. Sci.*, 407:189, 1998.

121. E. C. Walter, B. J. Murray, F. Favier, G. Kaltenpoth, M. Grunze, and R. M. Penner. Noble and coinage metal nanowires by electrochemical step edge decoration. *J. Phys. Chem. B*, 106:11407, 2002.

122. E. J. Menke, Q. Li, and R. M. Penner. Bismuth telluride (Bi2Te3) nanowires synthesized by cyclic electrodeposition/stripping coupled with step edge decoration. *Nano Letters*, 4:2009, 2004.

123. Frederic Juillerat, Harun H Solak, Paul Bowen, and Heinrich Hofmann. Fabrication of large-area ordered arrays of nanoparticles on patterned substrates. *Nanotechnology*, 16:1311, 2005.

124. D. Xia, A. Biswas, D. Li, and S.R.J. Brueck. Directed self-assembly of silica nanoparticles into nanometer-scale patterned surfaces using spin-coating. *Adv. Mat.*, 16:1427, 2004.

125. Pascale A. Maury, David N. Reinhoudt, and Jurriaan Huskens. Assembly of nanoparticles on patterned surfaces by noncovalent interactions. *Curr. Op. Coll. Int. Sci.*, 13:74, 2008.

126. Jinan Chai, Dong Wang, Xiangning Fan, and Jillian M. Buriak. Assembly of aligned linear metallic patterns on silicon. *Nat. Nano.*, 2:500, 2007.

127. R. Bennewitz. Structured surfaces of wide band gap insulators as templates for overgrowth of adsorbates. *J. Phys. Condens. Mat.*, 18:R417, 2006.

128. Jason R. Heffelfinger and C. Barry Carter. Mechanisms of surface faceting and coarsening. *Surf. Sci.*, 389:188, 1997.

129. L Nony, R Bennewitz, O Pfeiffer, E Gnecco, A Baratoff, E Meyer, T Eguchi, A Gourdon, and C Joachim. Cu-TBPP and PTCDA molecules on insulating surfaces studied by ultra-high-vacuum non-contact AFM. *Nanotechnology*, 15:S91, 2004.

130. R. C. Barrett and C. F. Quate. Imaging polished sapphire with atomic force microscopy. *J. Vac. Sci. Tech. A*, 8:400, 1990.

131. A. Sugawara and K. Mae. Surface morphology of epitaxial LiF(110) and CaF$_2$(110) layers. *J. Vac. Sci. Tech. B*, 23:443, 2005.

132. Akira Sugawara, T. Coyle, G. G. Hembree, and M. R. Scheinfein. Self-organized Fe nanowire arrays prepared by shadow deposition on NaCl(110) templates. *App. Phys. Lett.*, 70:1043, 1997.

133. A. Sugawara. Quasi-one-dimensional cobalt particle arrays embedded in 5 nm-wide gold nanowires. *IEEE Trans. Mag.*, 37:2123, 2001.

134. Takeshi Kitahara, Akira Sugawara, Haruyuki Sano, and Goro Mizutani. Anisotropic optical second-harmonic generation from the Au nanowire array on the NaCl(110) template. *App. Surf. Sci.*, 219:271, 2003.

Chapter 2
Theory

It is well known that light has the character of waves. Each electromagnetic (EM) wave propagates in space and time following the laws of electromagnetism, and consists of electric $\mathbf{E}(\mathbf{r}, t)$ and magnetic $\mathbf{B}(\mathbf{r}, t)$ fields oscillating in phase, perpendicular to each other and to the direction of propagation. The simplest form of EM wave is the *monochromatic plane wave*, as shown in Fig. 2.1, in which the $\mathbf{E}$ and $\mathbf{B}$ fields have the functional form of sinusoids, and that propagates along a fixed direction represented by its wave vector $\mathbf{k}$, with temporal period T and wavelength λ; using the complex notation, we can write

$$E(\mathbf{r}, t) = E \exp\left[i\left(\omega t - \mathbf{k} \cdot \mathbf{r}\right)\right] \tag{2.1a}$$

$$B(\mathbf{r}, t) = B \exp\left[i\left(\omega t - \mathbf{k} \cdot \mathbf{r}\right)\right] \tag{2.1b}$$

As plane waves, the EM fields have the same amplitude and the same phase on each plane perpendicular to the direction $\hat{\mathbf{k}}$. The angular frequency ω and the wave number $\mathbf{k}$ are defined as

$$\omega = \frac{2\pi}{T} \tag{2.2a}$$

$$\mathbf{k} = \frac{2\pi}{\lambda}\hat{\mathbf{k}} \tag{2.2b}$$

The travelling velocity is $s = \omega/k = \lambda/T$, and in vacuum or air has the constant value $c \approx 3 \cdot 10^8$ m/s independent on λ or T.

Real light beams are not monochromatic, but are instead the superimposition (*wavepacket*) of several EM waves, having different periods and wavelengths. In these cases, the dispersion curve $\omega(k)$ describes how the frequency of the EM component of the beam is related to its wave number k; in vacuum $\omega = ck$, while for propagation in dense media the relation is more complex. Since a wavepacket is composed of several waves propagating at different speeds, the term "wave velocity" is often ambiguous and can be defined in several ways. The *phase velocity* $v_p = \omega/k$ is the ordinary speed of any single component of the beam. The *group velocity*,

L. Anghinolfi, *Self-Organized Arrays of Gold Nanoparticles*, Springer Theses,
DOI: 10.1007/978-3-642-30496-5_2, © Springer-Verlag Berlin Heidelberg 2012

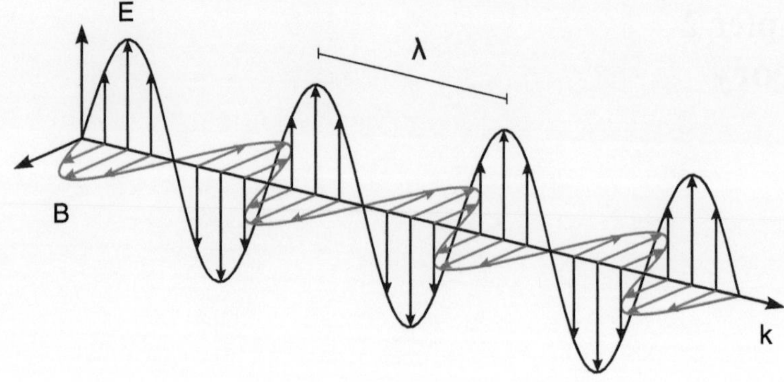

Fig. 2.1 Schematic representation of a plane monochromatic EM wave, linearly polarized, propagating along the direction defined by its wave vector **k**. The electric **E** and magnetic **B** fields are in the vertical and horizontal planes, respectively

defined as $v_g = \partial\omega/\partial k$, is instead the velocity with which the overall envelope of the wavepacket propagates.

Equations (2.1) define the amplitudes of the EM fields in a plane wave, but do not specify their direction of oscillation. In general, the EM fields of the waves composing a light beam can be randomly oriented (perpendicularly to the direction of propagation), in which case the beam is said *unpolarized*. In contrast, when the state of oscillation of the EM fields, called *polarization*, is the same for all the components of the beam, light is said *polarized*. For a EM plane wave, three possible states of polarization can be defined, *linear*, *circular* and *elliptical*. For linear polarized light (Fig. 2.2a) the orientation of the electric fields is constant along a specific direction, so that, as the wave propagates, they oscillate in a fixed plane called polarization plane. Instead, when light is elliptically (Fig. 2.2b) or circularly (Fig. 2.2c) polarized the electric field vector viewed along the propagation direction describes an ellipse (or a circle) around its wave vector **k**. The rotation is defined right-handed or left-handed when the observer sees the fields rotate counter-clockwise or clockwise, respectively.

2.1 Light and Matter

When light propagates inside a medium, its characteristics of propagation are modified by the interactions with the electrical charges of the material. From a microscopic point of view, each atom acts like a polarizable entity, which irradiates like a point dipole when excited by the oscillating electric field; the transmitted light is then determined by the superimposition of the incident radiation with the radiation emitted from all the atomic dipoles. Depending on the frequency of the exciting field, several mechanisms of polarization are possible (such as electric, atomic or orientational), and each of them is associated to a dielectric *polarizability* tensor $\boldsymbol{\alpha}$ (frequency dependent), defined through the relation

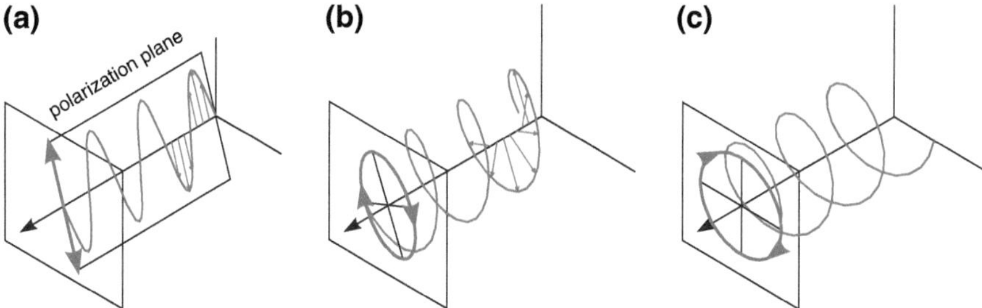

Fig. 2.2 Diagram of different states of polarization of light, and corresponding paths traced by the tip of the electric field vector (*red lines*). Panel **a** linear polarization, **E** oscillates along a constant direction. Panel **b** elliptical polarization, **E** traces an ellipse in each plane perpendicular the direction of propagation. Panel **c** circular polarization, special case of elliptical polarization where **E** traces a circle

$$\mathbf{p} = \varepsilon_0 \boldsymbol{\alpha} \otimes \mathbf{E} \tag{2.3}$$

where **p** is the induced electric dipole and **E** the exciting field; for isotropic polarization mechanisms, $\boldsymbol{\alpha}$ reduces to a constant and **p** and **E** are parallel. The *electric displacement field* **D** is defined as the sum of the exciting electric field and of the dipole density **P**, and is proportional to the *complex dielectric constant* (or *complex dielectric function*) $\varepsilon \equiv \varepsilon_1 - i\varepsilon_2$:

$$\mathbf{D} = \varepsilon_0 \mathbf{E} + \mathbf{P} = \varepsilon_0 \varepsilon \mathbf{E} \tag{2.4}$$

The complex dielectric function ε for a dense medium in general is a strong function of the radiation frequency, and contains all the information on the microscopic interactions between light and matter.

From the macroscopic point of view, the propagation of light in matter is characterized, in general, by a gradual decrease of the EM fields amplitude with the increasing distance travelled inside the dense medium (a phenomenon known as light *absorption*), and by the appearance of a frequency-dependent speed $s(\omega)$ of the propagating EM fields (a phenomenon known as light *dispersion*). These effects can be accounted for by the *complex refractive index* $\tilde{n}$, written as

$$\tilde{n} = N - iK \tag{2.5}$$

where

$$N = \frac{c}{s} \tag{2.6}$$

is the classic *refractive index* and K is the *extinction coefficient*.

The wave number inside a dense medium is redefined as

$$\mathbf{k} = \frac{\omega}{s}\hat{\mathbf{k}} = \frac{N\omega}{c}\hat{\mathbf{k}} \tag{2.7a}$$

or in the complex form

$$\tilde{\mathbf{k}} = \frac{\tilde{n}\omega}{c}\hat{\mathbf{k}} = \frac{\omega}{c}(N - iK)\,\hat{\mathbf{k}} \tag{2.7b}$$

The expressions (2.1) for the EM fields amplitudes then rewrite

$$E(\mathbf{r}, t) = E \exp\left[i\left(\omega t - \tilde{\mathbf{k}} \cdot \mathbf{r}\right)\right] \tag{2.8a}$$

$$B(\mathbf{r}, t) = B \exp\left[i\left(\omega t - \tilde{\mathbf{k}} \cdot \mathbf{r}\right)\right] \tag{2.8b}$$

from which, separating the real and imaginary part of the complex wave number, we obtain

$$\begin{aligned}
E(\mathbf{r}, t) &= E \exp\left(-\frac{\omega K}{c}\hat{\mathbf{k}} \cdot \mathbf{r}\right) \exp\left[i\left(\omega t - \mathbf{k} \cdot \mathbf{r}\right)\right] \\
&= E \exp\left(-\frac{\alpha}{2}\,\hat{\mathbf{k}} \cdot \mathbf{r}\right) \exp\left[i\left(\omega t - \mathbf{k} \cdot \mathbf{r}\right)\right]
\end{aligned} \tag{2.9}$$

The *absorption coefficient*

$$\alpha = \frac{2K\omega}{c} = \frac{4\pi K}{\lambda} \tag{2.10}$$

is defined as the fraction of power absorbed per unit length, as expressed by the *Beer law* $I(z+d) = I(z)\,e^{-\alpha d}$, where $I(z)$ and $I(z+d)$ are the intensities (optical power per unit area) at positions z and $z + d$. Then, we can see that the refractive index N and the extinction coefficient K are responsible, respectively, for the propagation of light and for the exponential decrease of the EM fields amplitudes.

For isotropic non-magnetic media, the complex refractive index and the dielectric constant are related by the simple expression

$$\varepsilon = \tilde{n}^2 \tag{2.11}$$

or equivalently for the real and imaginary parts

$$\varepsilon_1 = N^2 - K^2 \tag{2.12a}$$

$$\varepsilon_2 = 2NK \tag{2.12b}$$

and

$$N = \sqrt{\frac{\sqrt{\varepsilon_1^2 + \varepsilon_2^2} + \varepsilon_1}{2}} \tag{2.12c}$$

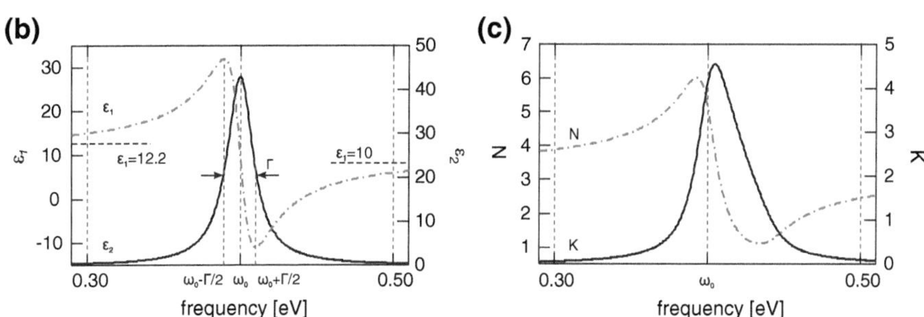

Fig. 2.3 *Top panel* physical representation of the Lorentz oscillator; when displaced by the application of an external electric field **E**, the positive and negative atomic charges attract each other by an elastic restoring force $\mathbf{F}_{el}$, and their motion is damped by a viscous force $\mathbf{F}_{\Gamma}$. *Bottom panels* frequency dependence of the real and imaginary parts of the complex dielectric function (*left panel*) and refractive index (*right panel*), calculated according to the Lorentz model (2.13) ($\chi = 9$, $A = 0.36$, $\omega_0 = 0.4\,\text{eV}$, $\Gamma = 20\,\text{meV}$) at frequencies close to resonance

$$K = \sqrt{\frac{\sqrt{\varepsilon_1^2 + \varepsilon_2^2} - \varepsilon_1}{2}} \tag{2.12d}$$

2.1.1 Dipole Oscillator Model

As seen before, the dielectric constant can be put in relation with the microscopic characteristics of the dense medium. In this section, we will therefore shortly discuss the main features of the microscopic mechanisms that govern the light-matter interactions. In general, the functional form of ε is quite complex, as several kinds of polarizations can be induced. When an analytical representation is required, the usual approach is therefore to decompose and analyse the individual contributions, and then merge the results. In this respect, several models have been proposed, suitable to describe the specific properties of the samples. The *dipole oscillator* or *Lorentz model* follows from the classical theory of absorption and despite its simplicity it offers a good picture of the polarization mechanisms.

According to this model, when an atom is irradiated by an external electric field it behaves like a damped harmonic oscillator (Fig. 2.3a): the exciting field displaces the positive nucleus from the negative electronic cloud, inducing an electric dipole; the charges, being separated, attract each other with a restoring force proportional to the displacement, realizing an oscillator; during the motion of the electrical charges under the influence of the fields, several energy losses can occur, like collisions with other atoms or spontaneous emission, effectively providing a damping mechanism for the oscillator.

The complex dielectric constant of a single Lorentz oscillator can be written as

$$\varepsilon^L(\omega) = 1 + \chi + \frac{A}{\omega_0^2 - \omega^2 + i\Gamma\omega} \tag{2.13}$$

where ω is the frequency of the exciting field, ω_0 the resonance frequency, Γ the damping rate, A a constant related to the electrons mass and density and χ is the susceptibility accounting for all the other contributions to the polarizability. The frequency dependence of the real and the imaginary parts of ε^L is plotted in Fig. 2.3b. ε_2^L is zero everywhere except near the resonance where a characteristic (*lorentzian*) peak is present, with full width at half maximum (FWHM) equal to Γ. ε_1^L, instead, has a more complex trend; at low frequencies it has a constant value of $1 + \chi + A/\omega_0^2$, then, approaching the resonance, it gradually rises up to a maximum at $\omega_0 - \Gamma/2$, it falls sharply to a minimum at $\omega_0 + \Gamma/2$, and it rises again, towards the high frequencies limit of $1 + \chi$.

Using the real and imaginary parts of (2.13), we can apply (2.12) to calculate the corresponding refractive index and extinction coefficient, as shown in Fig. 2.3c. Comparing Fig. 2.3b, c, we see that N is very similar to ε_1^L while K is peaked around $\approx \omega_0$ like ε_2^L. Indeed, if ε_2^L were much smaller than ε_1^L, it would follow $N \approx \sqrt{\varepsilon_1^L}$ and $K \approx \varepsilon_2^L/2\sqrt{\varepsilon_1^L}$. This correspondence, however, is only valid for gaseous phases, where the density of atoms is very low, while for solids it is only approximate because the absorptions are very strong; nevertheless, the absorption peak is generally observed at a frequency very close to ω_0.

Equation (2.13) can be generalized to include the contributions of several concomitant resonances occurring in the same medium at different frequencies, writing:

$$\varepsilon(\omega) = 1 + \sum_j \frac{A_j}{\omega_{0j}^2 - \omega^2 + i\Gamma_j\omega} \tag{2.14}$$

where j is the index numbering the (supposed discrete) oscillators of the medium. The overall frequency dependence of the dielectric constants according to Eq. (2.14) is schematically shown in Fig. 2.4.

This classical picture of the interaction between light and matter offers a simple physical interpretation of the dielectric constant. Looking at Fig. 2.4, we can identify parts where ε_1 is slowly varying and ε_2 is almost null, and regions where ε_1 rapidly changes and ε_2 has a maximum. Moreover, every time a resonance is crossed, moving from low to high frequencies, the average value of ε_1 decreases, approaching the value of 1 beyond the last oscillator. This can be understood in terms of polarizability of the material: considering a single mechanism of polarization, the atoms can follow the external field, i.e. the medium can be polarized, only up to the resonance fre-

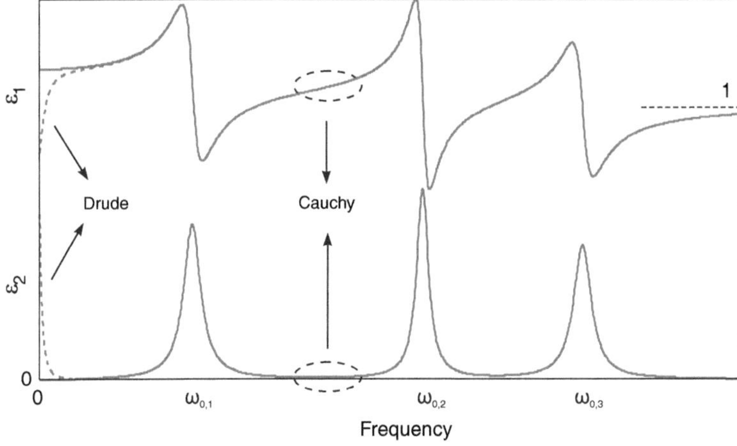

Fig. 2.4 Schematic diagram of the frequency dependence of the dielectric function of a hypothetical solid with three resonant frequencies ($\omega_{0,i}$). The Drude contribution is sketched in *dashed lines* at the lowest frequencies; an example of region of validity of the Cauchy parametrization is highlighted at the center of the figure

quency, where the polarization is maximum; at higher frequencies the field varies too fast and the average polarization reduces to zero. At very high frequencies, no polarization is more possible, and the material becomes completely transparent.

Drude Model for Free Electrons

The dipole oscillator model remains valid also to describe the free electrons in metals or the free carriers in semiconductors. These charges are not bound to the atoms nuclei, and therefore do not experience any restoring forces. Then, they can be treated as Lorentz oscillators with $\omega_0 = 0$, having dielectric constant (from (2.13))

$$\varepsilon(\omega) = 1 + \frac{A}{i\Gamma\omega - \omega^2}$$

This is usually rewritten in the form

$$\varepsilon_{Drude}(\omega) = 1 - \frac{\omega_P^2}{\omega^2 - i\Gamma\omega}, \tag{2.15}$$

known as the *Drude* dielectric function (Fig. 2.4).

According to the Drude-Sommerfeld model, the valence electrons in metals are considered as free particles, which do not interact with each other but can only undergo instantaneous collisions, with a characteristic scattering time $\tau = \Gamma^{-1}$. The sources of scattering can be various, for example impurities, defects or other electrons; if the mechanisms are independent from each other, the collision times sum up according to the Matthiessen's rule:

$$\frac{1}{\tau} = \sum_i \frac{1}{\tau_i} \tag{2.16}$$

The scattering time τ defines also a *mean free path* between subsequent collisions, given by $\lambda_{mfp} = v_F \tau$, where v_F is the Fermi velocity.

The quantity ω_P in (2.15) is known as the *plasma frequency*, and within the Drude model is given by

$$\omega_P = \sqrt{\frac{4\pi n e^2}{m}} \tag{2.17}$$

where n and m are the electron density and effective mass. It corresponds to the natural frequency of the free electrons charge density oscillations, i.e. periodic displacements of the free electrons gas as a whole. These excitations are called *plasmons*, and can be extended on the whole volume of the crystal or can be confined to the surface (*surface plasmons*). In addition, metallic structures with typical size of 1–10^2 nm can also sustain *localized surface plasmons*; they will be discussed in Sect. 2.3.

Sellmeier and Cauchy Models

Since in this thesis we will deal with transparent materials, we report here two useful approximations of the Lorentz model commonly used to describe the optical properties of materials in the transparent regions of the EM spectrum.

The *Sellmeier model* is derived from the Lorentz model (2.13) for $\omega \ll \omega_0$, assuming $\varepsilon_2 \approx 0$ and $\Gamma \to 0$. Rewriting (2.13) in terms of the wavelength ($\omega/c = 2\pi/\lambda$) we obtain

$$\varepsilon(\lambda) = \varepsilon_1(\lambda) = 1 + \frac{A}{(2\pi c)^2} \frac{\lambda_0^2 \lambda^2}{\lambda^2 - \lambda_0^2} \tag{2.18}$$

with $\lambda_0 = 2\pi c/\omega_0$. Then, the Sellmeier dielectric constant is written as

$$\varepsilon_1(\lambda) = N^2(\lambda) = A + \sum_j \frac{B_j \lambda^2}{\lambda^2 - \lambda_0^2}, \quad \varepsilon_2(\lambda) = 0 \tag{2.19}$$

where A and B_j are numerical parameters.

The *Cauchy model* is a further approximation of the Sellmeier model, obtained from the series expansion of (2.18):

$$N(\lambda) = A + \frac{B}{\lambda^2} + \frac{C}{\lambda^4} + \cdots, \quad K(\lambda) = 0 \tag{2.20}$$

Kramers-Kronig Relationships

We conclude this section with an important consideration on the relation between
the real and the imaginary parts of the complex dielectric function ε. Discussing
the Lorentz model, we have seen qualitatively that ε_1 and ε_2 are not independent
parameters but are related to each other. This is indeed a true property of ε, which
follows from the principle of causality. In fact, from the definition (2.4) we can write
the polarization of the medium as $P = \varepsilon_0(\varepsilon - 1)E$, which explicitly shows that
$\varepsilon(\omega) - 1$ is the response function for the application of electric fields. Therefore,
we can apply the laws of causality and derive the general *Kramers-Kronig* (KK)
relationships between the real and the imaginary parts of $\varepsilon(\omega)$ (but also of the complex
refractive index $\tilde{n}(\omega)$):

$$\varepsilon_1(\omega) = 1 + \frac{1}{\pi}\mathbf{P}\int_{-\infty}^{\infty}\frac{\varepsilon_2(\omega')}{\omega' - \omega}d\omega' \tag{2.21a}$$

$$\varepsilon_2(\omega) = -\frac{1}{\pi}\mathbf{P}\int_{-\infty}^{\infty}\frac{\varepsilon_1(\omega') - 1}{\omega' - \omega}d\omega' \tag{2.21b}$$

where $\mathbf{P}$ indicates the principal value of the integral.

The KK relations can be very useful because they allow, for example, to calculate
the dispersion of the dielectric constant and of the refractive index by measuring
the frequency dependence of only the optical absorption. They also provide a tool
for checking the physical consistency of the dielectric constant approximations. For
example, the Lorentz and Drude expressions for $\varepsilon(\omega)$ satisfy the KK relations, con-
trarily to the Sellmeier and Cauchy parametrizations, where $\varepsilon_2(\omega) = 0$ and $K(\lambda) = 0$
are not physically reasonable.

2.1.2 Light Refraction

When EM waves travel in homogeneous media they propagate according to (2.8),
maintaining constant direction, frequency and wavelength. Instead, when light
crosses different materials, it experiences a discontinuity of the refractive index,
which strongly affects its propagation. As a consequence of this two main effects
are observed (*refraction* of light): the transmitted beam propagates along a different
direction and with a different wavelength, and a reflected beam is generated. This is
illustrated in the following example.

Let us consider a flat interface between two media a and b with complex refractive
indices $\tilde{n}_a$ and $\tilde{n}_b$; an EM wave is approaching from a at an angle θ_i with respect
to the surface normal, while the reflected and the transmitted waves leave at angles
θ_r and θ_t (Fig. 2.5). (In the following the subscripts i, r and t will be used for the

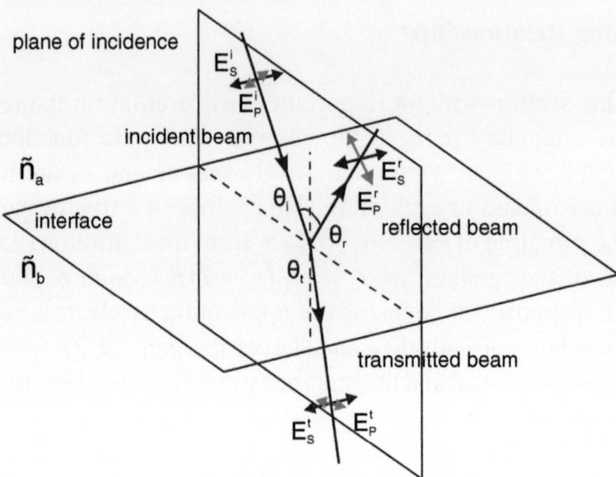

Fig. 2.5 Schematic representation of reflection and refraction of a monochromatic plane wave at the interface between two isotropic materials *a* and *b*

incident, the reflected and the transmitted beams, respectively). All the three beams are contained in the same plane, called the *plane of incidence*.

The refraction of light is usually computed imposing the continuity of the components of the EM fields at the interface, as follows from the Maxwell's equations. Requiring the equality of the phases ($\omega t - \mathbf{k} \cdot \mathbf{r}$) we obtain that the three beams have the same frequency ($\omega_i = \omega_r = \omega_t$), and that the reflected beam leaves at a direction specular to the incident one ($\theta_r = \theta_i$); the transmitted beam instead follows the *Snell law*:

$$N_a \sin \theta_i = N_b \sin \theta_t \tag{2.22}$$

The continuity of the amplitudes depends on the orientation of the EM fields, so usually two different directions are chosen as reference, the so-called *p* and *s*; they are defined, respectively, as the components of the EM fields parallel and perpendicular (*senkrecht* in German) to the plane of incidence, as sketched in Fig. 2.5. Employing this notation, we obtain the following *Fresnel coefficients*, defined as the ratios between the amplitude of the electric field associated to the reflected or transmitted beam to that associated to the incident beam:

$$r_s = \frac{E_p^r}{E_p^i} = \frac{\tilde{n}_a \cos \theta_i - \tilde{n}_b \cos \theta_r}{\tilde{n}_a \cos \theta_i + \tilde{n}_b \cos \theta_r} \tag{2.23a}$$

$$r_p = \frac{E_p^r}{E_p^i} = \frac{\tilde{n}_b \cos \theta_i - \tilde{n}_a \cos \theta_r}{\tilde{n}_b \cos \theta_i + \tilde{n}_a \cos \theta_r} \tag{2.23b}$$

$$t_s = \frac{E_s^t}{E_s^i} = \frac{2\, \tilde{n}_a \cos \theta_i}{\tilde{n}_a \cos \theta_i + \tilde{n}_b \cos \theta_r} \tag{2.23c}$$

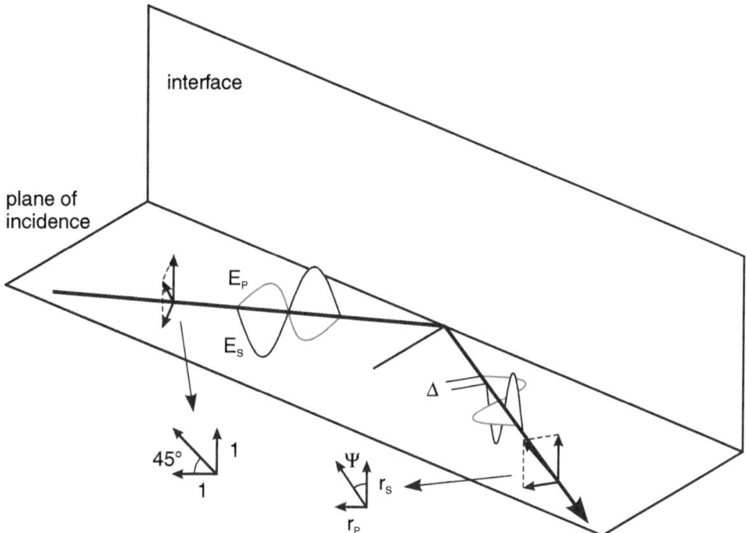

Fig. 2.6 An interpretation of Ψ and Δ upon reflection of a linearly polarized monochromatic beam at an interface between two media. The incident beam has two equally large s and p components. The p component experiences a reflection coefficient r_p, the s component a reflection coefficient r_s (adapted from Ref. [1])

$$t_p = \frac{E_p^t}{E_p^i} = \frac{2\,\tilde{n}_a \cos\theta_i}{\tilde{n}_b \cos\theta_i + \tilde{n}_a \cos\theta_r} \tag{2.23d}$$

These coefficients are complex numbers, because the refraction modifies both the amplitudes and the phases of the fields.

The corresponding relations for the intensities, called *reflectance* and *transmittance*, are given by

$$R_{p,s} = |r_{p,s}|^2 \tag{2.24a}$$

$$T_{p,s} = \frac{N_b \cos\theta_t}{N_a \cos\theta_i} |t_{p,s}|^2 \tag{2.24b}$$

As we will see later, other useful coefficients to describe the reflection of light from a surface are the so-called ellipsometric angles Ψ and Δ, defined through the polar form [2]

$$\rho = \tan\Psi e^{i\Delta} = \frac{r_p}{r_s} = \frac{|r_p|e^{i\delta_p}}{|r_s|e^{i\delta_s}} = \frac{|r_p|}{|r_s|}e^{i(\delta_p-\delta_s)} \tag{2.25}$$

From the previous definition it follows that $\tan\Psi$ is the ratio between the amplitudes of the p- and s-components of the reflected electric field; Δ instead is a more

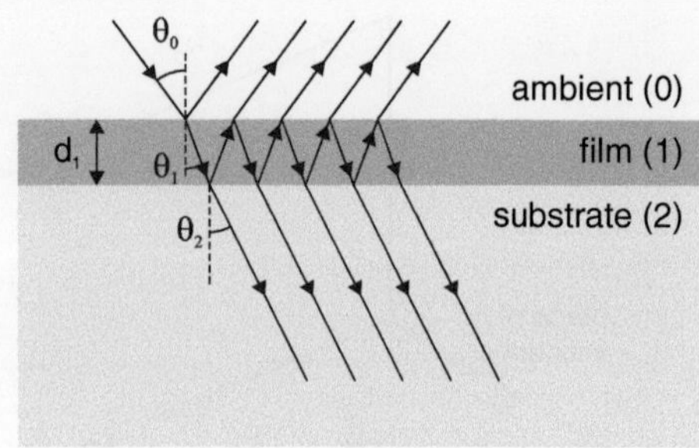

Fig. 2.7 Reflection and refraction of a light beam incident on a transparent thin film supported on a substrate

subtle parameter. In (2.25), δ_p and δ_s are the differences of the phases of the s and p components between the reflected and incident electric fields; Δ is the ulterior difference between these values. A simple graphical interpretation of these parameters is reported in Fig. 2.6, adapted from Ref. [1].

2.1.3 Reflection from Thin Films

A configuration of considerable importance for the optical measurements related to this work is the ambient/film/substrate system, composed of a thin film on top of a semi-infinite substrate. The thickness of the film is comparable with the wavelengths of interest, so that multiple reflections between the interfaces must be taken into account.

Here we consider the case of homogeneous and isotropic materials, sharply separated by flat boundaries parallel to each other (Fig. 2.7). The refractive indices of ambient, film and substrate are, respectively, $\tilde{n}_0$, $\tilde{n}_1$ and $\tilde{n}_2$, while d_1 is the thickness of the film and θ_0 the angle of incidence (and also of the beam transmitted beyond the substrate). In the contest of this work, the ambient will always be air ($\tilde{n}_0 \approx 1$) and the substrate a transparent medium, i.e. $\tilde{n}_2$ will be real.

The Fresnel coefficients for the reflection and transmission are given by [2]

$$r^\xi = \frac{r_{01}^\xi + r_{12}^\xi e^{-2i\beta}}{1 + r_{01}^\xi r_{12}^\xi e^{-2i\beta}} \tag{2.26a}$$

$$t^\xi = t_{20}^\xi \frac{t_{01}^\xi t_{12}^\xi e^{-i\beta}}{1 + r_{01}^\xi r_{12}^\xi e^{-2i\beta}} \qquad \xi = p, s \tag{2.26b}$$

where r_{hl} are the Fresnel coefficients (2.23) for the single hl interface, and

$$2\beta = 2\frac{2\pi d_1}{\lambda}\tilde{n}_1 \cos\theta_1 \tag{2.27}$$

is the phase shift acquired during a complete forward and backward reflection inside the film.

2.2 Heterogeneous Media

In the previous sections, we discussed the propagation of light inside homogeneous materials, and we saw that the optical response is completely resolved by the dielectric constant or the refractive index. It is not unusual, however, to deal with optical media that are formed by composites of several phases. In these cases it is well known that grain boundaries, voids, disordered regions or other inhomogeneities, on the length scale of several tens of nanometers, significantly affect the optical properties in the visible and near-UV range. In particular, screening charge developing at the grains boundaries and electrostatic interactions between adjacent grains considerably alter the local electric field and so the induced polarization; moreover, these effects depend on the shape and relative size of the microscopic structures.

Despite the extremely complicated microstructure of such composite media, in most cases it is still possible to describe the macroscopic response of such a heterogeneous material with an *effective* dielectric constant, that is an appropriate functional that "effectively" accounts for most of the optical characteristics of the specimens. In many cases, the effective dielectric constants of a mixture can be expressed in terms of a functional of the dielectric constants of each of the materials that enter the composite medium. Such approaches go under the name of "effective medium theories". There are several methods to derive effective medium theory; here, we start from the Clausius-Mossotti (CM) problem and generalize the solution to obtain the Lorentz–Lorenz, Maxwell-Garnett and Bruggeman expressions [3, 4].

The CM problem applies to a simple cubic lattice of polarizable points, with polarizability α and lattice constant a. When a uniform electric field $\mathbf{E}$ is applied, an electric dipole $\mathbf{p} = \varepsilon_0 \alpha \mathbf{E}_{loc}$ is induced at each lattice point $\mathbf{R}_n$, which in turn generates an electric field; the local field $\mathbf{E}_{loc}$ is therefore determined by the superimposition of the external field and the fields $\mathbf{E}_{dip}$ from the other dipoles:

$$\mathbf{E}_{loc}(\mathbf{r}) = \mathbf{E} + \sum_{\mathbf{R}_n} \mathbf{E}_{dip}(\mathbf{r} - \mathbf{R}_n) \tag{2.28}$$

where

$$\mathbf{E}_{dip}(\mathbf{r}) = \frac{1}{4\pi\varepsilon_0} \frac{3(\mathbf{p} \cdot \mathbf{r})\mathbf{r} - r^2\mathbf{p}}{r^5} \tag{2.29}$$

The macroscopic polarization $\mathbf{P}$ is defined as the average dipole moment per unit volume, and in this case reads

$$\mathbf{P} = \frac{N}{V}\varepsilon_0 \alpha \mathbf{E}_{loc} = n\varepsilon_0 \alpha \mathbf{E}_{loc} \tag{2.30}$$

where $n = a^{-3}$ is the volume density of points. In order to write $\mathbf{E}_{loc}$ as a function of $\mathbf{P}$ and $\mathbf{E}$ we need to evaluate the sum in (2.28). This can be done using the Lorentz cavity method. In the simplest approximation we consider a sphere of radius ρ centered in $\mathbf{r}$: we explicitly add the dipole fields from the points inside the sphere and average the dipoles outside. Given the cubic symmetry of the lattice, the former contribution vanishes, while the latter equals to the volume integral of a dipole $\mathbf{P}$, which is $\mathbf{P}/3\varepsilon_0$ [5]. Then, Eq. (2.30) becomes

$$\mathbf{P} = n\varepsilon_0\alpha\left(\mathbf{E} + \frac{\mathbf{P}}{3\varepsilon_0}\right) \tag{2.31}$$

and we obtain for $\mathbf{P}$

$$\mathbf{P} = \varepsilon_0\frac{n\alpha}{1 - n\alpha/3}\mathbf{E} \tag{2.32}$$

Finally, applying the definition of the dielectric constant

$$\mathbf{D} = \varepsilon_0\varepsilon\mathbf{E} = \varepsilon_0\mathbf{E} + \mathbf{P} \tag{2.33}$$

we found the CM relation

$$\frac{\varepsilon - 1}{\varepsilon + 2} = \frac{n\alpha}{3} \tag{2.34}$$

Now, the simplest heterogeneous medium can be realized by randomly assigning to the points of the preceding system two different polarizabilities α_a and α_b. In this case we find

$$\frac{\varepsilon - 1}{\varepsilon + 2} = \frac{n_a\alpha_a}{3} + \frac{n_b\alpha_b}{3} \tag{2.35}$$

In this expression ε represents the effective dielectric constant of the composite. For practical uses, this form is not very useful, because it contains the microstructural parameters n_i and α_i, which are difficult to measure. Instead, if the dielectric constants ε_a and ε_b of the pure phases are available, we can use Eq. (2.34) to rewrite Eq. (2.35) as

$$\frac{\varepsilon - 1}{\varepsilon + 2} = f_a\frac{\varepsilon_a - 1}{\varepsilon_a + 2} + f_b\frac{\varepsilon_b - 1}{\varepsilon_b + 2} \tag{2.36}$$

where $f_{a,b} = n_{a,b}/(n_a + n_b)$ are the volume fractions of the two phases. This is the *Lorentz–Lorenz* effective medium expression.

In more realistic situations, however, the phases a and b are not uniformly mixed at atomic scale, but rather form grains large enough to possess their own dielectric identity.[1] Repeating the same derivation as before, we now identify these grains as polarizable entities, so we cannot assume anymore they are immersed in vacuum. If we suppose each entity immersed in a host with dielectric constant ε_h, Eq. (2.36) becomes

$$\frac{\varepsilon - \varepsilon_h}{\varepsilon + 2\varepsilon_h} = f_a\frac{\varepsilon_a - \varepsilon_h}{\varepsilon_a + 2\varepsilon_h} + f_b\frac{\varepsilon_b - \varepsilon_h}{\varepsilon_b + 2\varepsilon_h} \tag{2.37}$$

---

[1] The grains must contains an enough number of atoms to develop their characteristic band structure.

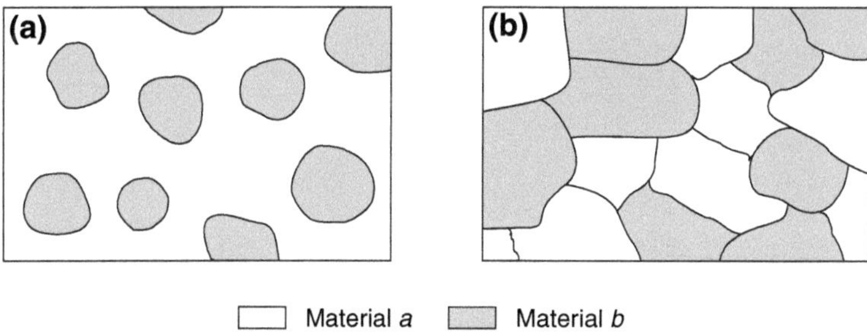

Fig. 2.8 Schematic representation of the microstructure of two different heterogeneous two-phases media. Panel **a** separate grains of material A dispersed in a continuous host of material B (suitable for Maxwell-Garnett EMA (2.38)). Panel **b** random mixture of grains of the two constituents (suitable for Bruggemann EMA (2.39))

This form is still somewhat incomplete, because the dependence of ε_h on ε_a and ε_b is unspecified.

Then, let's suppose that b is a dilute phase inside a, i.e. $f_b \ll f_a$: we can choose $\varepsilon_h \approx \varepsilon_a$, from which we obtain

$$\frac{\varepsilon^{MG} - \varepsilon_a}{\varepsilon^{MG} + 2\,\varepsilon_a} = f_b \frac{\varepsilon_b - \varepsilon_a}{\varepsilon_b + 2\,\varepsilon_a} \tag{2.38}$$

This and the equivalent equation for $f_a \ll f_b$ are the *Maxwell-Garnett* effective medium equations [6, 7]. They are usually applied when dealing with systems composed of well isolated particles or grains dispersed in continuous media (Fig. 2.8a).

In cases where f_a and f_b are comparable, it may not be clear the distinction between host and inclusions. Then, we can make the self-consistent choice $\varepsilon = \varepsilon_h$, and (2.37) reduces to

$$0 = f_a \frac{\varepsilon_a - \varepsilon^{Br}}{\varepsilon_a + 2\,\varepsilon^{Br}} + f_b \frac{\varepsilon_b - \varepsilon^{Br}}{\varepsilon_b + 2\,\varepsilon^{Br}} \tag{2.39}$$

This is the *Bruggeman* expression [8], commonly known as *effective medium approximation* (EMA) and suited to describe uniform mixtures of two different materials (Fig. 2.8b).

The above effective medium expressions are few of the simplest approximations for describing the optical constants of heterogeneous media. One of the main assumptions is that all the phases feel the same equivalent mean field, implying that the domains are uniformly distributed in the volume of the medium. This is not the case, for example, when the grains are coherently arranged on a lattice [9, 10], so that the local field distribution has the same symmetry of the lattice, or are strongly coupled by EM radiation [11, 12], so that the local field is highly localized (see Sect. 2.3). Another situation where such EMAs fail is the proximity of percolation, when long-range conductive paths are established between the grains of the single

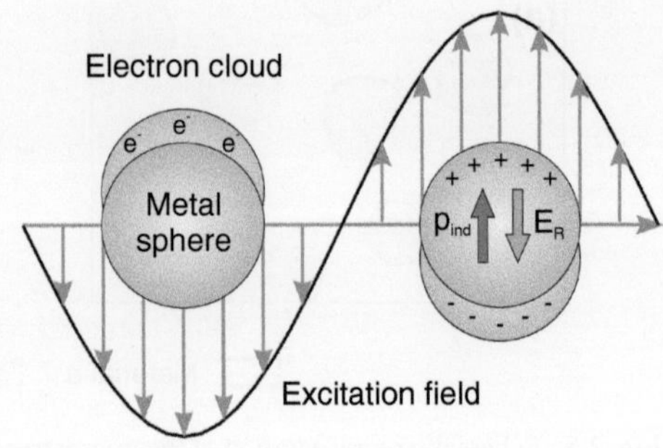

Fig. 2.9 Sketch of homogeneous metallic spheres placed in a oscillating EM field. The conduction electrons are displaced as a whole, polarizing the sphere ($\mathbf{p}_{ind}$), while the surface of the particles exerts a restoring force $\mathbf{E}_R$, so that resonance conditions can be established, leading to EM field amplification inside and in proximity of the particle

phases [13–15]. The application of effective medium theories must be therefore carefully evaluated depending on the specific case under scrutiny, taking into account that they are always a simplification of heterogeneous systems into equivalent single-phase media.

2.3 Optical Properties of Metallic Nanostructures

The optical properties of metals, from low energies up to the near-UV, are mostly dominated by the contribution of the free electrons, which are weakly bound to the metallic atoms and can freely move inside the crystal. These electrons, for example, are responsible for the high electric conductivity and the high optical reflectivity and absorption from DC up to visible and near UV frequencies.

On the other hand, metallic nanostructures with sub-micrometer dimensions exhibit very different optical responses with respect to their bulk counterparts. An external EM field can penetrate inside the volume of the particles, shifting the free electrons gas with respect to the ions lattice; consequently, charges of opposite sign accumulate on the opposite surfaces of the particles, polarizing the metal and establishing restoring local fields ($\mathbf{E}_R$ in Fig. 2.9). Therefore, in formal analogy with the Lorentz model, the particles can be viewed as oscillators, whose behaviour is determined by the free electrons effective mass, charge and density, but most importantly by the geometry of the particles [16–20]. Under resonance conditions, the free electrons gas is coherently dragged by the external excitation, so the electric dipoles induced inside each particles become extremely large. Correspondingly, the local fields in proximity of the particles are order of magnitudes enhanced with respect to the incident fields, the scattering cross section is enormously amplified, and very strong absorption peaks are observed. Such collective excitations are commonly known as *localized surface plasmons* (LSPs) [16–18, 21–23]; for "common" metals they are usually observed in the visible range (Ag [24, 25], Au [23] or Cu [26, 27]) and deep UV (Al [28, 29]).

In general, the optical response of metal nanoparticles can be quite complex, as the particles have more than a single resonant mode. These modes differ in their

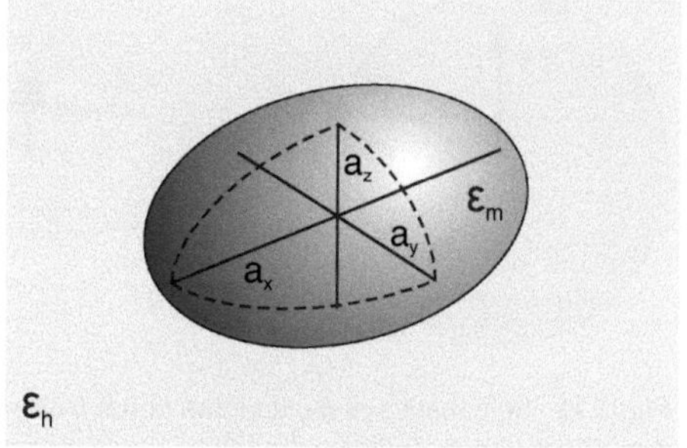

Fig. 2.10 Sketch of a isolated metallic ellipsoidal particle, with principal semiaxis (a_x, a_y, a_z) and dielectric function ε_m, immersed in a dielectric host of dielectric constant ε_h

charge and field distribution, and are strongly dependent on the particle's size (with respect to the EM wavelength), shape and environment. The analytical treatment of LSPs for particles with arbitrary shape is therefore almost always not feasible, and computational methods are required. Indeed, only few simple configurations allow the exact solution of the optical response, which include spherical particles [30, 31], spheroids [32] and infinite long cylinders [33].

The *Mie* theory [30] is an exact solution of the Maxwell equations for the scattering and absorption problem of spherical particles, and it is usually employed to derive approximate solutions for similar geometries. According to this theory, the EM fields are expanded in spherical harmonics, and all the possible LSP modes correspond to the dipolar and multipolar EM eigenmodes of the particle. A full treatment of the EM interaction within the Mie theory is however a very challenging task, because the analytical description of the highest polar modes is very complex. Therefore, the Mie theory is often approximated to include only the most significant contributions. The excitation strength of each mode is determined by the corresponding expansion of the EM field; in particular, when the particles are much smaller than the involved wavelengths (typically up to tens of nanometers for EM fields in the visible range) the resonances are mainly dipolar in character, so only the first order terms can be retained. In such cases the Mie solution reduces to the *Rayleigh* approximation for the elastic scattering of light [31], and the quasi-static approximation can be invoked to apply the equations of electrostatics in electromagnetism.

2.3.1 Quasi-Static Approximation

For particles whose size is small compared to local variations of the incident light, the phase of the EM fields varies very little over the particles volume and we can assume uniform and non-retarded fields: this is called the *quasi-static approximation* (QSA). For common metals, like Ag, Au, Cu, Al, which have the LSP resonances in the visible and UV range, this approximation can adequately describe the optical response of spherical and ellipsoidal particles with sizes below ≈ 100 nm.

Let's consider a metallic ellipsoidal particle immersed in a transparent dielectric host and far from any other polarizable entity (Fig. 2.10). The ellipsoid has semiaxes

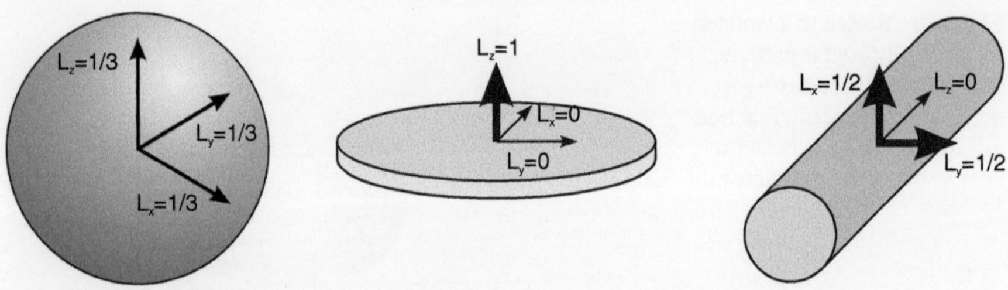

Fig. 2.11 Schematics of the depolarization factors for different shapes of homogeneous nanoparticles. *Black arrows* indicate the relative strengths of the induced electric dipoles along the principal axes. Panel **a** spherical particle, $a_x = a_y = a_z$. Panel **b** disc, $a_x, a_y \gg a_z$. Panel **c** cylinder, $a_x, a_y \ll a_z$

a_γ ($\gamma = x, y, z$) oriented along the cartesian axes, and the dielectric constants of the metal and the host are, respectively, ε_m and ε_h (the latter purely real). We start by considering the effects of a static applied electric field $\mathbf{E}_0$. As the particle is not spherical, the polarizability is a tensor $\boldsymbol{\alpha}$. Then, in general the induced electric dipole $\mathbf{p}$ is not parallel to $\mathbf{E}_0$ [5, 31]:

$$\mathbf{p} = \varepsilon_0 \boldsymbol{\alpha} \otimes \mathbf{E}_{loc} \tag{2.40}$$

where $\mathbf{E}_{loc}$ is the local field acting on the particle, and differs from $\mathbf{E}_0$ due to the polarization of the host. If the host is homogeneous and isotropic, then $\mathbf{E}_{loc} = \varepsilon_h \mathbf{E}_0$, and the previous equation becomes

$$\mathbf{p} = \varepsilon_0 \varepsilon_h \boldsymbol{\alpha} \otimes \mathbf{E}_0 \tag{2.41}$$

For ellipsoidal particles $\boldsymbol{\alpha}$ is diagonal, and has principal values α_γ given by [31]

$$\alpha_\gamma = v \frac{\varepsilon_m - \varepsilon_h}{\varepsilon_h + L_\gamma(\varepsilon_m - \varepsilon_h)} \quad \gamma = x, y, z \tag{2.42}$$

where $v = 4\pi/3 \, a_x a_y a_z$ is the particle's volume and L_γ are the *depolarization factors*, that can be written as

$$L_\gamma = \frac{a_x a_y a_z}{2} \int_0^\infty \frac{dq}{(q + a_\gamma^2)\sqrt{\prod_{\eta=x,y,z}(q + a_\eta^2)}} \tag{2.43}$$

and satisfy the sum rule $\sum_\gamma L_\gamma = 1$.

For spherical particles (Fig. 2.11a) we have $a_x = a_y = a_z \equiv a$ and the geometrical factors reduce to $L_\gamma = 1/3$ in all directions: the polarizability is isotropic and from (2.42) we find

$$\alpha_{\text{sph}} = 3 \, v \frac{\varepsilon_m - \varepsilon_h}{\varepsilon_m + 2 \, \varepsilon_h} \tag{2.44}$$

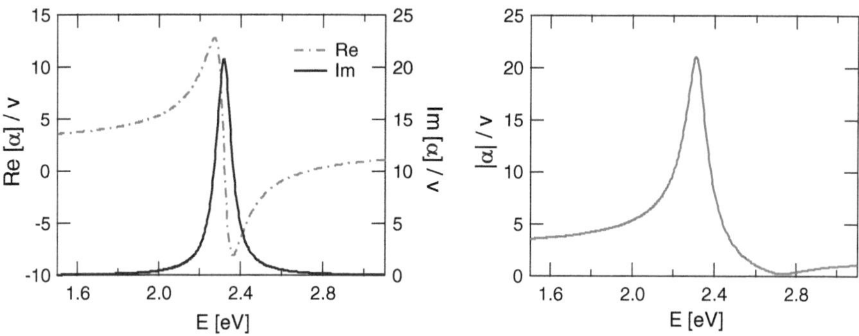

Fig. 2.12 Real (*red line*) and imaginary (*black line*) parts (*left panel*) and magnitude (*right panel*) of the complex polarizability α for a gold sphere of radius $a = 30\,\mathrm{nm}$, immersed in a homogeneous host with dielectric constant $\varepsilon_h = 1.4$. The complex dielectric function of gold is shown in Fig. 2.13

Two other interesting cases are the limits to disks and to cylinders. A disk is obtained when the ellipsoid is considerably stretched along two of the principal axes, or equivalently shrunk along one of them, so that, for example, $a_x, a_y \gg a_z$ (Fig. 2.11b). In such case the depolarization factors in the plane of the disk reduce to 0, while the one along the minor axis grows up to 1; correspondingly, no polarization becomes possible within the plane, and the disk can be polarized only along the normal axis. In the opposite case, i.e. $a_z \gg a_x, a_y$, the ellipsoid evolves instead to a cylinder (Fig. 2.11c). Now, the depolarization factor along the main axis (L_z) becomes 0, implying that the cylinder can be polarized only by electric fields lying in the xy plane.

In Fig. 2.12 the real and imaginary parts and the absolute value of α_{sph} are reported for a metallic sphere in air. We can clearly see a strong enhancement of the polarizability, corresponding to a minimum of $|\varepsilon_m + 2\varepsilon_h|$. The magnitude of α_{sph} at resonance does not diverge, but it is limited due to the imaginary part of the dielectric constant. If $\mathrm{Im}\,[\varepsilon_m(\omega)]$ is small or slowly-varying, the resonance condition simplifies to

$$\mathrm{Re}\,[\varepsilon_m(\omega)] = -2\,\varepsilon_h \tag{2.45}$$

which is called the *Fröhlich condition* for the LSPs resonances.

The total electric field outside the particle is the sum of the incident field and the dipolar field generated by the particle,

$$\mathbf{E}(\mathbf{r}) = \mathbf{E}_0 + \frac{1}{4\pi\varepsilon_0\varepsilon_h}\frac{1}{r^3}\frac{3(\mathbf{r}\cdot\mathbf{p})\mathbf{r} - r^2\mathbf{p}}{r^2} \tag{2.46}$$

from which we can see that the resonances of $\boldsymbol{\alpha}$ (and $\mathbf{p}$) also determine resonant enhancements of $\mathbf{E}$.

Given the solution for electrostatics, we can now turn our attention to EM fields. In the quasi-static regime we are dealing with particles much smaller than the wavelengths, i.e. $a_\gamma \ll \lambda$, so we can consider time varying fields and neglect spatial retardation effects. If we assume an incident plane wave radiation, the exciting electric field is given by $\mathbf{E}_{\mathrm{ex}}(\mathbf{r}, t) = \mathbf{E}_0 e^{i\omega t}$ and induces a time-varying dipole moment

$$\mathbf{p}(t) = \varepsilon_0 \varepsilon_h \boldsymbol{\alpha} \otimes \mathbf{E}_0 \, e^{i\omega t}, \qquad (2.47)$$

This oscillating dipole irradiates in the surrounding space, leading to the scattering of the incident plane wave. The dipole fields are now given by [5]

$$\mathbf{H}(\mathbf{r}, t) = \frac{c}{4\pi} \frac{1}{r^3} \left[(kr)^2 + ikr \right] \frac{\mathbf{r} \times \mathbf{p}}{r} e^{i(\omega t - kr)} \qquad (2.48a)$$

$$\mathbf{E}(\mathbf{r}, t) = \frac{1}{4\pi \varepsilon_0 \varepsilon_h} \frac{1}{r^3} \left[(kr)^2 \frac{(\mathbf{r} \times \mathbf{p}) \times \mathbf{r}}{r^2} + (1 - ikr) \frac{3(\mathbf{r} \cdot \mathbf{p})\mathbf{r} - r^2 \mathbf{p}}{r^2} \right] e^{i(\omega t - kr)}$$
$$(2.48b)$$

In particular, we can identify two limiting spatial domains. A *near field* component dominates in the vicinity of the particle ($kr \ll 1$) and decays from the particle center proportionally to r^{-3}; in this regime the electrostatic result (2.46) is recovered for the electric field (with the additional exponential time dependence), while the magnetic field reduces to

$$\mathbf{H}(\mathbf{r}, t) = \frac{ic}{4\pi} \frac{kr}{r^3} \frac{\mathbf{r} \times \mathbf{p}}{r} e^{i\omega t} \qquad (2.49)$$

Then, in the near field regime the retardation effects can be neglected, and the fields are predominantly electric, as the magnitude of the magnetic field is about a factor $\varepsilon_0 c \, kr$ smaller than that of the electric field.

The other limit is the *far field* regime, acting at distances much larger than the wavelengths ($kr \gg 1$). In this regime the fields are proportional to r^{-1} and have the form of spherical waves:

$$\mathbf{H}(\mathbf{r}, t) = \frac{ck^2}{4\pi} \frac{\mathbf{r} \times \mathbf{p}}{r} \frac{e^{i(\omega t - kr)}}{r} \qquad (2.50a)$$

$$\mathbf{E}(\mathbf{r}, t) = \frac{c}{\varepsilon_0 \varepsilon_m} \frac{\mathbf{H} \times \mathbf{r}}{r} \qquad (2.50b)$$

Another consequence of the resonantly enhanced polarizability is the concomitant enhancement of the efficiency of the particle scattering and absorption. Within the quasi-static approximation, the corresponding cross-sections σ are given by the Rayleigh expressions [31]

$$\sigma_{\mathrm{sca}, \gamma} = \frac{k^4}{6\pi} |\alpha_\gamma|^2 \qquad (2.51a)$$

$$\sigma_{\text{abs},\gamma} = k \, \text{Im}\left[\alpha_\gamma\right] \qquad (2.51\text{b})$$

As the polarizability is proportional to the volume, we can see that σ_{sca} depends on the square of the volume while σ_{abs} scales only linearly with v. Therefore, small particles prevalently absorb light, while the scattering process is dominant in large particles. We note that in the derivation of (2.51) no explicit assumptions on the dielectric constants are made, so they are valid also for dielectric scatterers. In such a case, they demonstrate a very crucial problem for optical measurements of ensembles of nanoparticles: due to the rapid scaling of the scattering cross-section, $\sigma_{\text{sca}} \propto a^6$, it is very difficult to pick out small objects from a background of larger scatterers.

2.3.2 Beyond the Quasi-Static Approximation

Despite its simplicity, we can see from the polarizabilities (2.42) that the quasi-static theory already accounts for the main effects associated with the major parameters affecting the LSPs resonances, i.e. the influence of particle shape and size, the metal and the environment optical characteristics. However, comparing the experimental results with the QSA predictions some inconsistencies remain, mainly related to the linewidth of the resonances and the influence of the particle size. Remaining in the dipolar modes regime, we can introduce two corrections to the QSA, which account for surface damping in particles with dimensions smaller than the mean free path of the oscillating electrons, and retardation effects in larger particles.

Surface Damping

For very small metallic nanoparticles, with sizes comparable to the electrons mean free path λ_{mfp}, the bulk dielectric constant is modified by the additional scattering of the electrons at the particle surfaces. This *surface damping* destroys the coherent oscillations of the electrons, resulting in a broadening of the LSP resonances. For common metals λ_{mfp} is usually of the order of 30–50 nm, so the scattering is dominant for dimensions below $\approx$20 nm.

To account for these finite size effects, we start from the dielectric constant $\varepsilon_{\text{exp}}(\omega)$ measured experimentally for the bulk metal. This can be decomposed in contributions from interband transitions, between states separated by an energy gap, and intraband transitions, between states at the Fermi level in incompletely filled bands:

$$\varepsilon_{\text{exp}}(\omega) = \varepsilon_{\text{inter}}(\omega) + \varepsilon_{\text{intra}}(\omega) \qquad (2.52)$$

Due to the presence of the gap, the former have features at high energies, usually starting from the near-UV range. On the contrary, intraband transitions are promoted by low-energy photons and involve quasi-free electrons at the Fermi level. Then, for $\varepsilon_{\text{intra}}$ we can employ the Drude theory (2.15):

$$\varepsilon_{\text{intra}}(\omega) = 1 - \frac{\omega_P^2}{\omega^2 - i\Gamma_0\omega} \tag{2.53}$$

where $\Gamma_0 = v_F/\lambda_{\text{mfp}}$ is determined by the electrons mean free path in the metal bulk (v_F is the Fermi velocity). Surface damping can be empirically modeled [34] as an additional size-dependent contribution $\Gamma_{\text{surf}}(a)$ to Γ_0, which writes

$$\Gamma_0(a) = \Gamma_0 + \Gamma_{\text{surf}}(a) = \frac{v_F}{\lambda_{\text{mfp}}} + A\frac{v_F}{a}. \tag{2.54}$$

A is an empirical factor, of the order of 1, which incorporates the details of the scattering processes [21]; for a sphere it is usually chosen between 3/4 and 1 [16, 18, 35, 36]. Introducing Γ_0 in $\varepsilon_{\text{intra}}$, we obtain

$$\varepsilon_{\text{intra}}(\omega, a) = 1 - \frac{\omega_P^2}{\omega^2 - i\Gamma_0(a)\omega} \tag{2.55}$$

Now, we can modify the dielectric constant by subtracting from (2.52) the bulk intraband contribution (2.53) and adding the corrected term (2.55): we find the size-dependent $\varepsilon_m(\omega, a)$ given by

$$\begin{aligned}
\varepsilon_m(\omega, a) &= \varepsilon_{\text{exp}}(\omega) - \varepsilon_{\text{intra}}(\omega) + \varepsilon_{\text{intra}}(\omega, a) \\
&= \varepsilon_{\text{exp}}(\omega) + \Delta\varepsilon(\omega, a)
\end{aligned} \tag{2.56}$$

with [37]

$$\Delta\varepsilon(\omega, a) = \frac{\omega_P^2}{\omega}\left(\frac{1}{\omega - i\Gamma_0} - \frac{1}{\omega - i\Gamma_0(a)}\right) \tag{2.57}$$

In Fig. 2.13, the dielectric constants of gold nanoparticles with radius $a = 5$ nm and 50 nm are compared. We can see that reducing the particle size the imaginary part of ε_m increases at the larger wavelengths (lower energies), while the real part is only slightly raised. Then, according to the Fröhlich condition (2.45), the resonances do not experience significant shifts, and the main effects of surface damping is a broadening of the plasmons linewidth, in accordance with the experimental results [21].

Retardation Effects

In the previous sections we have seen that in the quasi-static approximation the incident EM radiation is considered uniform within the particles volume. This assumption can be adequate for particles with sizes up to 100 nm and EM frequencies in the visible range, however it fails to predict the dependence of the optical response on the NP dimensions. Moreover, as the variations of the incident EM fields cannot be neglected anymore, multipolar eigenmodes are also excited. Nevertheless, if the NPs sizes are lower than $\approx 10\%$ of the typical wavelength of the incident radiation [38] (≈ 40 nm in the visible range), multipolar contributions can still be neglected

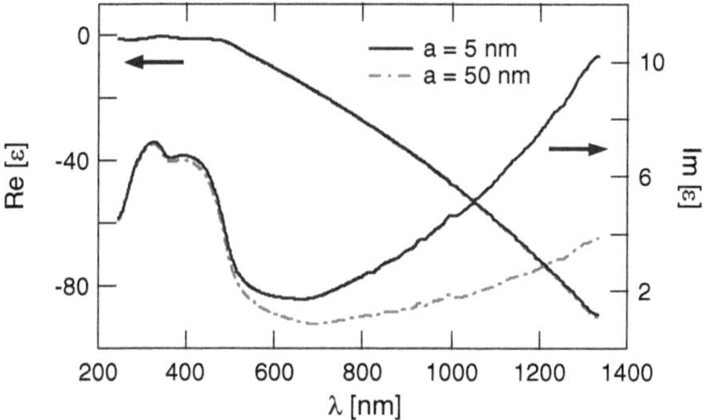

Fig. 2.13 Real and imaginary parts of the dielectric constant of gold spheres with different radii, computed applying the finite size corrections of Eq. (2.57). The optical constants of the bulk gold were extracted from ellipsometric measurements performed on a Au(111) crystal (Au/mica manufactured by Phases)

and retardation effects can be explicitly calculated for the dipolar modes [39]; this is sometimes called *modified long-wavelength approximation* (MLWA). For gold ellipsoidal nanoparticles, it has been shown that it consistently reproduces exact numerical solutions for particles with equivalent volumes up to a $\approx$40 nm radius sphere and aspect ratios below 10 [19, 40, 41], and it is still in qualitative agreement with results at $\approx$ 200 nm dimensions [42]. As the particles under scrutiny in this thesis have typical sizes of several tens of nanometers, we will also employ MLWA for calculating their polarizabilities.

Let's consider a spherical particle with radius a excited by an EM field. In QSA we found that the induced dipole **p** is proportional to the incident electric field (2.47). The electrodynamic corrections associated with MLWA are introduced by rewriting (2.47) as [39]

$$\mathbf{p} = \varepsilon_0 \varepsilon_h \alpha (\mathbf{E} + \mathbf{E}_{\mathrm{rad}}) \tag{2.58}$$

where the radiative correction field $\mathbf{E}_{\mathrm{rad}}$ is given by

$$\mathbf{E}_{\mathrm{rad}} = \frac{1}{4\pi \varepsilon_0 \varepsilon_h} \left(\frac{k^2}{a} + i\frac{2}{3}k^3 \right) \mathbf{p} \tag{2.59}$$

The first term in (2.59) comes from the depolarization of the radiation across the particle surface, due to the finite ratio of particle size to wavelength; the main effect of this *dynamic depolarization* is red shifting the plasmon resonance as the particle size is increased. The second term is the *radiation damping* due to the radiative losses of the induced dipole; it grows rapidly with particle size, reducing the intensity of the resonances and making the spectrum broader and asymmetric.

The net effect of these terms is to modify the total induced polarization, so that the polarizability of the sphere now rewrites as

$$\alpha_{\text{sph}}^{\text{MLWA}} = \frac{\alpha_{\text{sph}}}{1 - \dfrac{\alpha_{\text{sph}}}{4\pi}\left(\dfrac{k^2}{a} + \dfrac{2}{3}ik^3\right)} \tag{2.60}$$

where α_{sph} is the QSA expression (2.44). For ellipsoidal particles, excited along one of the principal axes, (2.60) remains valid substituting α_{sph} and a with the corresponding polarizability and semi-axis [19, 43].

2.3.3 Far Field Extinction Spectra of Nanoparticles Ensembles

In the previous sections the LSPs were reviewed within the quasi-static approximation, and we saw that many different factors affects the resonances, related both to the geometry of the system and to the dielectric environment. On the other hand, one of the key points of the current thesis is the fabrication and the detailed analysis of the plasmonic response of several arrays of gold NPs, characterized by different morphological and geometrical parameters. It is therefore very instructive, in order to better understand the experimental results and analysis, to present here some optical calculations for ideal ensembles of metallic nanoparticles, illustrating the behaviour of the LSP resonances under different system configurations.

We consider an ensemble of *non-interacting* metallic nanoparticles immersed in a homogeneous dielectric host. The latter has a purely real dielectric constant ε_h, while for the particles we employ a simple Drude model with finite-size corrections (2.55), thereby neglecting the contributions of the interband transitions; the physical constants have been chosen to fit the optical constants of gold [44]: $v_F = 1.4 \times 10^6$ m/s, $\omega_P = 9\,$eV, $\Gamma_0 = 70\,$meV. For these parameters, we present the calculated spectra for the extinction efficiencies Q_{ext}, defined as the sum of the absorption and scattering cross-sections (2.51) renormalized by the geometrical cross-sections πa_γ^2, as a function of the wavelength:

$$Q_{\text{ext},\gamma} = (\sigma_{\text{abs},\gamma} + \sigma_{\text{sca},\gamma})/\pi a_\gamma^2 \tag{2.61}$$

Influence of the Environment

We start by considering spherical particles of radius $a = 30\,$nm, and analyse the effects of a variation of the host dielectric constant ε_h. Since the particles are immersed in a dense medium, the local electric field differs from the external excitation due to the polarization $\mathbf{P}_h$ of the medium. Furthermore the particles itself are sources of electric fields, which modify $\mathbf{P}_h$, and therefore, again, the local field.

In Fig. 2.14 we report the calculated curves for Q_{ext} with ε_h varying from 1 (vacuum) up to 3, computed without applying the MLWA correction. We can see that the position of the resonance moves to larger wavelengths (lower energies) at higher ε_h and correspondingly its magnitude increases. Employing the Fröhlich condition

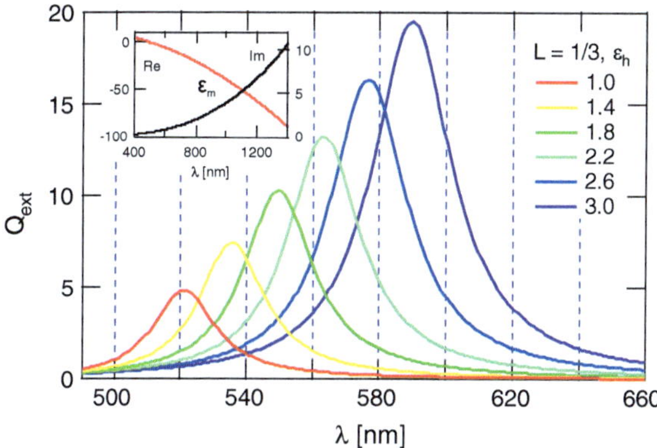

Fig. 2.14 Computed extinction efficiencies Q_{ext} for non-interacting gold spherical particles, of radius $a = 30\,$nm, as a function of the dielectric constant ε_h of the host. The particles' dielectric function, employed for the calculations, is plotted in the inset of the figure

(2.45) and looking at ε_m in the inset of Fig. 2.14, we can deduce that the resonance red-shift is due to the negative slope of the real part of ε_m; the enhancement of the resonances is instead related to the proportionality between the induced dipole and ε_h (see (2.41)), which at the resonance is about $p \propto \varepsilon_h^2$.

Influence of the Particle Shape

Now we set the dielectric constant of the host assuming, for example, $\varepsilon_h = 1.4$, and inspect the effects of the geometrical parameters on the extinction efficiency. In the case of a sphere, that we choose as a starting point, the isotropic particle shape leads to LSP resonances that are independent on the direction of the incident electric field. If we suppose to deform the sphere stretching it along one of its diameters, we obtain a so-called prolate spheroid, that is an ellipsoid with semi-axes $a_x = a_y < a_z$. The dielectric response becomes anisotropic depending on the orientation of the field with respect to the symmetry axis a_z, so that a so-called longitudinal (L) and a so-called transverse (T) mode can be identified, corresponding to excitation along the directions parallel or perpendicular to a_z. Stretching further the particle, these modes shift in frequency as a function of the ellipsoid aspect ratio, the longitudinal one red-shifting and the transverse one slightly blue-shifting. In general, for more anisotropic shapes, a greater number of different modes appear [18].

In Fig. 2.15 computed spectra for aspect ratios $a_z : a_{x,y}$ between $1 : 1$ and $5 : 1$ are shown. To understood the LSP shifts we look again at the resonance condition, which now becomes (cfr. Eq. (2.42))

$$\mathrm{Re}\left[\varepsilon_m(\omega)\right] \approx -\frac{1 - L_\gamma}{L_\gamma}\varepsilon_h. \tag{2.62}$$

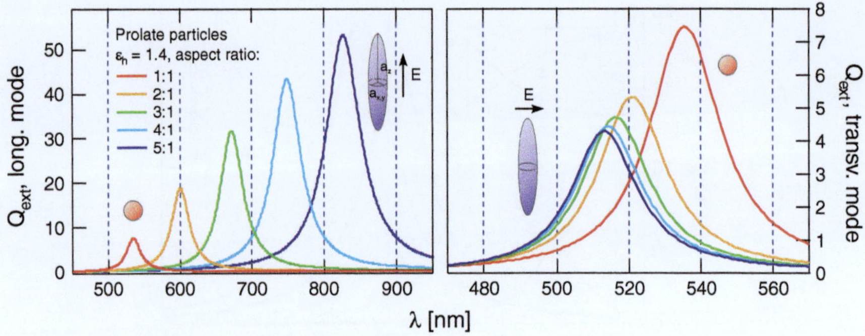

Fig. 2.15 Computed extinction efficiencies Q_{ext} for non-interacting gold prolate particles, having different aspect ratios, dispersed in a continuous homogeneous host ($\varepsilon_h = 1.4$). Electric field applied along (*left panel*) and transverse to (*right panel*) the long axis. The particles dielectric function is plotted in the inset of Fig. 2.14

For prolate geometries, the transversal depolarization factors L_T are larger than the longitudinal L_L, therefore, increasing the asymmetry of the particle, the ratio in (2.62) becomes smaller for the T modes and larger for the L modes; moreover, under resonance conditions, the polarizability is roughly proportional to $(1 - 2L_\gamma)/L_\gamma \varepsilon_h^2$. From these considerations, we can deduce that, in analogy with the previous case, the T (L) spectra are shifted towards high (low) energies, and the magnitudes are weakened (enhanced).

Influence of the Particle Size

The last parameter that we will consider in this analysis, which affects the LSP resonances, is the particle size. In Fig. 2.16a the extinction spectra, calculated employing the MLWA, are reported for spherical particles with radius between 5 and 75 nm; the corresponding peak positions and linewidths are highlighted in Fig. 2.16b. We can see that increasing the radius a the resonances are systematically red shifted, while their linewidth initially decreases, reaches a minimum at about 20 nm, and then monotonically increases. These trends are mainly due to surface damping and retardation effects. For small particles, the former is dominant, so the LSP position is slightly affected by the size, while the linewidth has a contribution $\Gamma(a) \propto a^{-1}$ (see Eq. (2.54)). At higher size, dynamic depolarization and radiation damping, respectively proportional to a^2 and a^3, rapidly grow, inducing a strong red shift and broadening of the peaks, and reducing their intensity.

Lastly, in Fig. 2.17 the absorption and scattering contributions to the total extinction efficiency are shown separately for two different particles with radius $a = 5$ and 50 nm, respectively. As already predicted before, in the former case Q_{ext} is entirely determined by absorption ($Q_{\text{abs}} \propto a$), the scattering efficiency being more than three orders of magnitude smaller. Increasing the volume, Q_{sca} grows much more rapidly

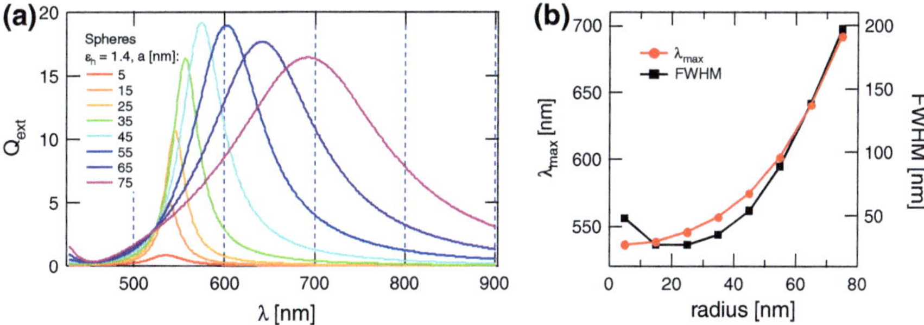

Fig. 2.16 *Left panel* computed extinction efficiencies Q_{ext} for non-interacting gold spherical particles, having different radii, dispersed in a homogeneous dielectric host ($\varepsilon_h = 1.4$); the dielectric function of gold is plotted in the inset of Fig. 2.14. *Right panel* position (λ_{max}, red line) and full width at half maximum (FWHM, *black curve*) of the LSP peaks in panel (**a**) as a function of the particle radius

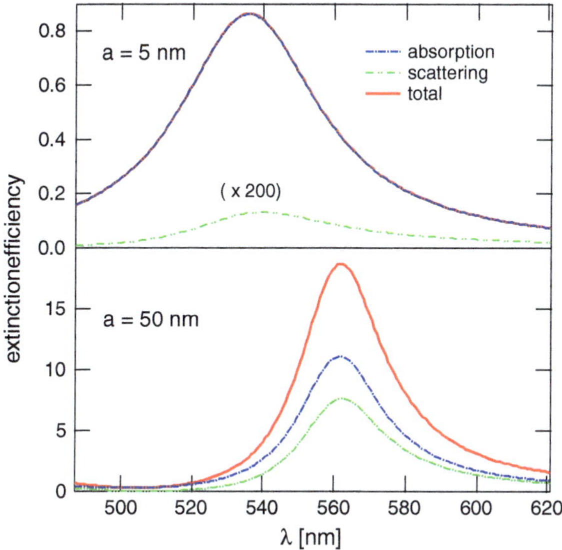

Fig. 2.17 Computed absorption (*blue curves*), scattering (*green curves*) and total (*red curves*) extinction efficiencies for gold spheres of radius $a = 5\,\text{nm}$ (*top panel*) and $a = 50\,\text{nm}$ (*bottom panel*) dispersed in a dielectric host ($\varepsilon_h = 1.4$). The particles dielectric function is plotted in the inset of Fig. 2.14

than Q_{abs} ($Q_{\text{sca}} \propto a^4$), for $a = 50\,\text{nm}$ is comparable with Q_{abs} and eventually it becomes dominant at higher radii.

2.3.4 Effects of EM Interactions

Up to this point we always considered ensembles of isolated nanoparticles, where the particles spacing was so high that the direct interactions could be neglected. When several particles are brought close to one another, electromagnetic coupling effects set in, and the LSP resonances are determined by the collective behaviour of the system, so that the overall optical response can be significantly modified with respect to the isolated case. Usually, two different interactions regimes are distinguished, depending on the interparticle distance d: for closely spaced particles, $d \ll \lambda$, near-field interactions proportional to d^{-3} dominate and the particles can be described as an array of interacting point dipoles; due to the rapid scaling of the interaction strength with distance, this regime holds for particle separations below ≈ 150 nm. For larger separations, the near field can be neglected and the dipolar coupling is mainly through far field of the scattered light, which scales as d^{-1}. In this work we will not consider the latter case, but only concentrate on the so-called near-field. In particular, in order to gain qualitative insight on the effects of interactions, we will analyse two different configurations, namely an isolated particle supported on a dielectric substrate and a pair of close particles.

Substrate Effects

In this section we consider a spherical metallic nanoparticle in the near vicinity of a flat substrate. When the particle is polarized by an incident EM wave, the irradiated field induces a distribution of charges in the substrate, which in turn modifies the local field acting on the particle. In general, this substrate-induced field is not homogeneous in space, so multipolar modes of higher order can be excited in addition to the dipolar one (Fig. 2.18a). However, if the particle is not in direct contact with the substrate but at least few nanometers away (Fig. 2.18b), the spatial variations of the local field are less pronounced and the QSA is still suited to describe the particle optical response [18].

Within the QSA, the induced distribution of charges can be seen as the particle image charge; then we can treat the polarization of the substrate like a point dipole $\mathbf{p}_{\mathrm{ind}}$ specular to the particle with respect to the substrate surface (Fig. 2.18). According to the electrostatic theory, $\mathbf{p}_{\mathrm{ind}}$ is given by [5]

$$\mathbf{p}_{\mathrm{ind}} = \frac{\varepsilon_s - \varepsilon_h}{\varepsilon_s + \varepsilon_h}(-p_x, -p_y, p_z) \tag{2.63}$$

where $\mathbf{p}$ is the particle dipole, ε_s and ε_h the dielectric constant of the substrate and the host.

Due to the presence of the substrate, the symmetry of the system is now broken and different plasmon modes can be found in the directions parallel or normal to the interface. To illustrate this, first let's suppose the external electric field $\mathbf{E}_{\mathrm{ext}}$ perpen-

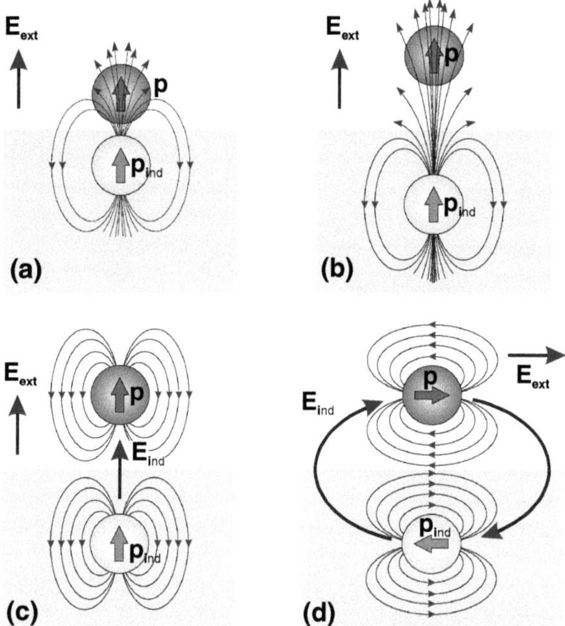

Fig. 2.18 *Top panels* metallic sphere in proximity of an interface between two different media, and electric field applied normal to the interface. When the sphere is nearly touching the interface, the electric field generated by the image charge is highly inhomogeneous within the particle's volume (panel **a**), and becomes more uniform as the particle moves away from the interface (panel **b**). *Bottom panels* image charge dipole, and corresponding dipolar field $\mathbf{E}_{ind}$ generated at the particle center, for excitation $\mathbf{E}_{ext}$ normal (panel **c**) and parallel (panel **d**) the interface. In both cases $\mathbf{E}_{ind}$ has the same direction of $\mathbf{E}_{ext}$

dicular the plane of the substrate (Fig. 2.18c): both the particle and the image charge get polarized in the same direction of $\mathbf{E}_{ext}$; then, the electric field $\mathbf{E}_{ind}$, generated by $\mathbf{p}_{ind}$ at the particle position, has the same orientation like $\mathbf{E}_{ext}$ and acts against the restoring forces of $\mathbf{p}$. Recalling that the particle can be modelled as an oscillator, the lowering of the restoring force is reflected by a reduction of the resonance frequency, i.e. a red-shift of the LSP. The same effect occurs when the $\mathbf{E}_{ext}$ is oriented in-plane (Fig. 2.18d): in this case $\mathbf{p}$ and $\mathbf{p}_{ind}$ are in the same direction of the field but antiparallel with respect to each other, so that $\mathbf{E}_{ind}$ is again directed like $\mathbf{E}_{ext}$.

Particle Interactions

We now discuss the effects on the LSP induced by near field coupling in a system of close particles. Common ensemble of nanoparticles include linear chains [46–50] and bidimensional arrays [51–56], however for a basic understanding of the

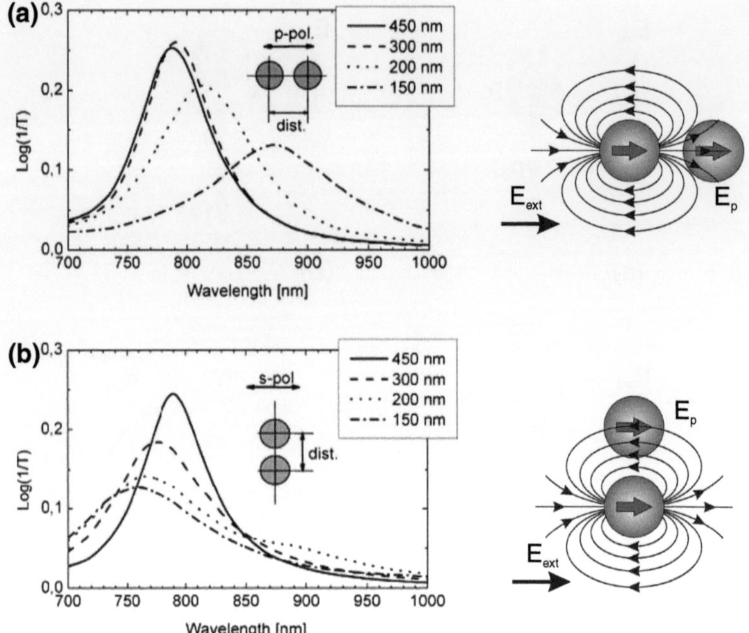

Fig. 2.19 Extinction (=log(1/Transmission)) spectra of 2D arrays of Au nanoparticles pairs (reprinted from Ref. [45], Copyright 2003, with permission from Elsevier), for different interparticle center-to-center distances. The polarization direction of the exciting light is **a** parallel to the long particle pair axis and **b** orthogonal to it

effects of interactions, we consider the simplest case of only two coupled particles [45, 57, 58].

In analogy to the previous configuration, if the particles are not too close to each other, we can employ the QSA and treat them as two interacting point dipoles. Again, the single particle symmetry is broken, and depending on the polarization with respect to the pair axis, we can distinguish a longitudinal (L) and a transverse (T) polarizations, respectively parallel and perpendicular the pair axis.

In Fig. 2.19a, c we report the experimental extinction spectra from Ref. [45], measured on arrays of pairs of nanodiscs for different interparticles spacings. When the external electric field is oriented along the pairs axis, the near field $\mathbf{E}_p$ generated by each particle on its companion is parallel to the polarization and acts against the restoring forces (Fig. 2.19b); then, similarly to the previous case, the longitudinal LSP is shifted at lower energies (Fig. 2.19a). On the contrary, if the electric field is perpendicular to the axis, $\mathbf{E}_p$ points in the opposite direction of the polarization, reducing the strength of the applied field or equivalently reinforcing the restoring forces (Fig. 2.19d); therefore, in this case the resonance frequency is increased, and the LPS mode blue-shifts (Fig. 2.19c). Moreover, due to the retardation effects of

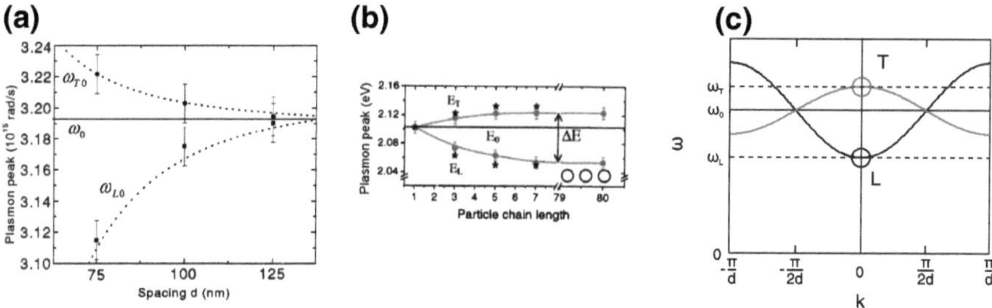

Fig. 2.20 Panel **a** dependence of the plasmon peak position on the interparticle spacing d for both the longitudinal (L) and transverse (T) excitation of the collective mode in 2D arrays of gold nanoparticles chains (reprinted with permission from Ref. [47]. Copyright 2002 by the American Physical Society); the *dotted line* shows the $1/d^3$ dependence predicted by a point dipole interaction model [61]. Panel **b** collective plasmon resonance energies for both L and T excitations for Au nanoparticle chains of different lengths (Reprinted with permission from Ref. [60]. Copyright 2002, American Institute of Physics). Panel **c** computed dispersion relation $\omega(k)$ for L and T LSP modes in a infinite chain of metallic nanoparticles, according to a point dipole interaction model [61, 62]

the interactions, the radiation damping of coupled particles is larger than of isolated ones, so the LSP modes become broader.

The splitting and broadening of the single-particle resonances are determined by the strength of the interactions between adjacent particles, and they can be tuned by varying the separation (Fig. 2.20a) or the number (Fig. 2.20b) of the particles forming the chain [36, 46, 47, 59, 60].

In particular, the splitting of the single-particle LSP modes has been demonstrated [61, 62] to play a key role for the coherent transport of EM energy. Exploiting the ability of metallic nanostructures to localize and strongly enhance the EM radiation, uniaxial chains of nanoparticles can be employed as waveguides at nanoscale, allowing the propagation of EM signals with lateral confinement beyond the diffraction limit [62]. All the main characteristics of the waveguide (group velocity, bandwidth, attenuation of the propagating waves) are dependent on the energy separation between the longitudinal and the transverse LSP modes, excited by electric fields along and normal the waveguide. For example, in Fig. 2.20c we report the dispersion relation $\omega(k)$ for the longitudinal L (blue line) and transverse T (red line) modes, computed for an infinite chain of spherical nanoparticles assuming dipolar interactions between nearest-neighbours [61, 62]; the group velocity v_g for energy transport corresponds to the slope of the curves. For both L and T modes, v_g is null and the frequency splitting $\Delta\omega$ is maximum for uniform excitations, i.e. $k = 0$ along the chain. Increasing the wave number k of the exciting field, v_g increases while $\Delta\omega$ decreases. For $k = \pi/d$ the splitting of the L and T modes collapses to zero and their energy reduces to the single-particle LSP; correspondingly, the group velocity reaches the maximum values and $v_g^L = 2v_g^T$, due to the stronger EM coupling for L waves than for T waves. Maximum values of v_g up to $0.1c$ and transmission efficiencies up to 64 % were predicted [61].

In summary, the basic theory of the interaction between light and heterogeneous media has been reviewed, with particular emphasis on the optical response of metallic inclusions embedded in dielectric hosts. The main feature of interest in view of the current work is the resonant behaviour of metallic nanoparticles under the excitation of an external EM radiation, which leads to a very strong amplification of the EM fields inside and in the near field range outside the particles. Correspondingly, such systems exhibit strong resonance peaks in the reflection and absorption of light, whose characteristics (position and linewidth) depend on both intrinsic geometrical factors (single-particle size and shape) and extrinsic parameters (dielectric constant of the host, proximity of a surface or other polarizable entities). Such features, not observed in the bulk counterparts, are a characteristic effect of the collective oscillations of the electrons gas confined inside the metallic structures, and are given the name of localized surface plasmon resonances.

The observation and analysis of the LSP resonances in 2D arrays of gold nanoparticles will be the main topic of the current thesis. The arguments presented in this chapter will be recalled in Sect. 6.2.1 to formulate an effective medium theory describing the optical response of the Au NPs arrays, and to provide a theoretical support to the experimental data.

References

1. J.H. den Boer. *Spectroscopic Infrared Ellipsometry: Components, Calibration and Applications*. PhD thesis, Eindhoven University of Technology, 1995.
2. H. Fujiwara. *Spectroscopic Ellipsometry. Principles and Applications*. Wiley, 2007.
3. G. A. Niklasson, C. G. Granqvist, and O. Hunderi. Effective medium models for the optical properties of inhomogeneous materials. *Appl. Opt.*, 20:26, 1981.
4. D.E. Aspnes. Optical properties of thin films. *Thin Solid Films*, 89:249, 1982.
5. J.D. Jackson. *Classical Electrodynamics*. John Wiley & Sons, 1999.
6. J.C.M. Garnett. Colours in metal glasses and in metallic films. *Philos. Trans. R. Soc. London.*, 203:385, 1904.
7. J.C.M. Garnett. Colours in metal glasses, in metallic films, and in metallic solutions. ii. *Philos. Trans. R. Soc. London.*, 205:237, 1906.
8. D.A.G. Bruggeman. Berechnung verschiedener physikalischer konstanten von heterogenen substanzen. *Ann. Phys. Leipzig*, 24:636, 1935.
9. F. J. García-Vidal, J. M. Pitarke, and J. B. Pendry. Effective medium theory of the optical properties of aligned carbon nanotubes. *Phys. Rev. Lett.*, 78:4289, 1997.
10. B. Sareni, L. Krähenbühl, A. Beroual, and C. Brosseau. Effective dielectric constant of periodic composite materials. *J. App. Phys.*, 80:1688, 1996.
11. Cheng-ping Huang, Xiao-gang Yin, Qian-jin Wang, Huang Huang, and Yong-yuan Zhu. Long-wavelength optical properties of a plasmonic crystal. *Phys. Rev. Lett.*, 104:016402, 2010.
12. V. G. Kravets, S. Neubeck, A. N. Grigorenko, and A. F. Kravets. Plasmonic blackbody: Strong absorption of light by metal nanoparticles embedded in a dielectric matrix. *Phys. Rev. B*, 81(16):165401, 2010.
13. F. Brouers, J. P. Clerc, G. Giraud, J. M. Laugier, and Z. A. Randriamantany. Dielectric and optical properties close to the percolation threshold. II. *Phys. Rev. B*, 47:666, 1993.
14. G. Polatsek and O. Entin-Wohlman. Effective-medium approximation for a percolation network: The structure factor and the Ioffe-Regel criterion. *Phys. Rev. B*, 37:7726, 1988.

15. Z.-M. Dang, Y. Shen, and C.-W. Nan. Dielectric behavior of three-phase percolative Ni-BaTiO$_3$/polyvinylidene fluoride composites. *App. Phys. Lett.*, 81:4814, 2002.
16. Poelsema B. Stefan Kooij E. Shape and size effects in the optical properties of metallic nanorods. *Phys. Chem. Chem. Phys.*, 8:3349, 2006.
17. S. Link, M. B. Mohamed, and M. A. El-Sayed. Simulation of the optical absorption spectra of gold nanorods as a function of their aspect ratio and the effect of the medium dielectric constant. *J. Phys. Chem. B*, 103:3073, 1999.
18. Cecilia Noguez. Surface plasmons on metal nanoparticles: The influence of shape and physical environment. *J. Phys. Chem. C*, 111:3806, 2007.
19. K. Lance Kelly, Eduardo Coronado, Lin Lin Zhao, and George C. Schatz. The optical properties of metal nanoparticles: The influence of size, shape, and dielectric environment. *J. Phys. Chem. B*, 107:668, 2003.
20. J. J. Mock, M. Barbic, D. R. Smith, D. A. Schultz, and S. Schultz. Shape effects in plasmon resonance of individual colloidal silver nanoparticles. *J. Chem. Phys.*, 116:6755, 2002.
21. U. Kreibig. Optical absorption of small metallic particles. *Surf. Sci.*, 156:678, 1985.
22. E. Hutter and J. H. Fendler. Exploitation of localized surface plasmon resonance. *Adv. Mater.*, 16:1685, 2004.
23. Stephan Link and Mostafa A. El-Sayed. Size and temperature dependence of the plasmon absorption of colloidal gold nanoparticles. *J. Phys. Chem. B*, 103:4212, 1999.
24. Paul Mulvaney. Surface plasmon spectroscopy of nanosized metal particles. *Langmuir*, 12:788, 1996.
25. Arnim Henglein and Michael Giersig. Formation of colloidal silver nanoparticles: Capping action of citrate. *J. Phys. Chem. B*, 103:9533, 1999.
26. George H. Chan, Jing Zhao, Erin M. Hicks, George C. Schatz, and Richard P. Van Duyne. Plasmonic properties of copper nanoparticles fabricated by nanosphere lithography. *Nano Letters*, 7:1947, 2007.
27. Manjeet Singh, I. Sinha, M. Premkumar, A.K. Singh, and R.K. Mandal. Structural and surface plasmon behavior of Cu nanoparticles using different stabilizers. *Coll. Surf. A*, 359:88, 2010.
28. P.R. West, S. Ishii, G.V. Naik, N.K. Emani, V.M. Shalaev, and A. Boltasseva. Searching for better plasmonic materials. *Laser Photon. Rev.*, 4:795, 2010.
29. Y. Ekinci, H. H. Solak, and J. F. Löffler. Plasmon resonances of aluminum nanoparticles and nanorods. *J. App. Phys.*, 104:083107, 2008.
30. G. Mie. Optical absorption spectra for silver spherical particles. *Ann. Phys.*, 25:377, 1908.
31. Bohren C.F.; Huffman D.R. *Absorption and scattering of light by small particles.* Wiley, 1998.
32. S. Asano; G. Yamamoto. Light scattering by a spheroidal particle. *Appl. Opt.*, 14:29, 1980.
33. A.C. Lind; J. M. Greenberg. Electromagnetic scattering by obliquely oriented cylinders. *J. Appl. Phys.*, 37:3195, 1966.
34. H. Hövel, S. Fritz, A. Hilger, U. Kreibig, and M. Vollmer. Width of cluster plasmon resonances: bulk dielectric functions and chemical interface damping. *Phys. Rev. B*, 48:18178, 1993.
35. U Kreibig. Electronic properties of small silver particles: the optical constants and their temperature dependence. *J. Phys. F: Met. Phys.*, 4:999, 1974.
36. Chan C.T. Fung K.H. A computational study of the optical response of strongly coupled metal nanoparticle chains. *Opt. Comm.*, 281:855, 2008.
37. F. Bisio, M. Palombo, M. Prato, O. Cavalleri, E. Barborini, S. Vinati, M. Franchi, L. Mattera, and M. Canepa. Optical properties of cluster-assembled nanoporous gold films. *Phys. Rev. B*, 80:205428, 2009.
38. Foss Jr. CA. Feldheim DL. *Metal nanoparticles: synthesis, characterization, and applications.* New York: Marcel Dekker, 2002.
39. M. Meier and A. Wokaun. Enhanced fields on large metal particles: dynamic depolarization. *Opt. Lett.*, 8:581, 1983.
40. Cecilia Noguez. Optical properties of isolated and supported metal nanoparticles. *Opt. Mat.*, 27:1204, 2005.
41. Kaspar D. Ko and Kimani C. Toussaint Jr. A simple gui for modeling the optical properties of single metal nanoparticles. *J. Quant. Spectr. Rad. Trans.*, 110:1037, 2009.

42. E.J. Zeman and G.C. Schatz. An accurate electromagnetic theory study of surface enhancement factors for Ag, Au, Cu, Li, Na, Al, Ga, In, Zn, and Cd. *J. Phys. Chem.*, 91:634, 1987.

43. Linda Gunnarsson, Tomas Rindzevicius, Juris Prikulis, Bengt Kasemo, Mikael Kll, Shengli Zou, and George C. Schatz. Confined plasmons in nanofabricated single silver particle pairs: Experimental observations of strong interparticle interactions. *J. Phys. Chem. B*, 109:1079, 2005.

44. P. Tamarat S. Berciaud, L. Cognet and B. Lounis. Observation of intrinsic size effects in the optical response of individual gold nanoparticles. *Nano Letters*, 5:515, 2005.

45. W. Rechberger, A. Hohenau, A. Leitner, J. R. Krenn, B. Lamprecht, and F. R. Aussenegg. Optical properties of two interacting gold nanoparticles. *Opt. Comm.*, 220:137, 2003.

46. Q.H. Wei, K.H. Su, S. Durant, and X. Zhang. Plasmon resonance of finite one-dimensional Au nanoparticle chains. *Nano Letters*, 4:1067, 2004.

47. Stefan A. Maier, Mark L. Brongersma, Pieter G. Kik, and Harry A. Atwater. Observation of near-field coupling in metal nanoparticle chains using far-field polarization spectroscopy. *Phys. Rev. B*, 65:193408, 2002.

48. Tian Yang and Kenneth B. Crozier. Dispersion and extinction of surface plasmons in an array of gold nanoparticle chains: influence of the air/glass interface. *Opt. Express*, 16:8570, 2008.

49. Yu Chen and A. M. Goldman. Formation of one-dimensional nanoparticle chains. *App. Phys. Lett.*, 91:063119, 2007.

50. Jiwen Zheng, Zihua Zhu, Haifeng Chen, and Zhongfan Liu. Nanopatterned assembling of colloidal gold nanoparticles on silicon. *Langmuir*, 16:4409, 2000.

51. Andrew N. Shipway, Eugenii Katz, and Itamar Willner. Nanoparticle arrays on surfaces for electronic, optical, and sensor applications. *ChemPhysChem*, 1:18, 2000.

52. Beomseok Kim, Steven L. Tripp, and Alexander Wei. Self-organization of large gold nanoparticle arrays. *J. Am. Chem. Soc.*, 123:7955, 2001.

53. N. Félidj, J. Aubard, G. Lévi, J. R. Krenn, A. Hohenau, G. Schider, A. Leitner, and F. R. Aussenegg. Optimized surface-enhanced Raman scattering on gold nanoparticle arrays. *Appl. Phys. Lett.*, 82:3095, 2003.

54. Christy L. Haynes and Richard P. Van Duyne. Nanosphere lithography: A versatile nanofabrication tool for studies of size-dependent nanoparticle optics. *J. Phys. Chem. B*, 105:5599, 2001.

55. Christy L. Haynes, Adam D. McFarland, LinLin Zhao, Richard P. Van Duyne, George C. Schatz, Linda Gunnarsson, Juris Prikulis, Bengt Kasemo, and Mikael Käll. Nanoparticle optics: The importance of radiative dipole coupling in two-dimensional nanoparticle arrays. *J. Phys. Chem. B*, 107:7337, 2003.

56. Marisa Mäder, Thomas Höche, Jürgen W. Gerlach, Susanne Perlt, Jens Dorfmüller, Michael Saliba, Ralf Vogelgesang, Klaus Kern, and Bernd Rauschenbach. Plasmonic activity of large-area gold nanodot arrays on arbitrary substrates. *Nano Letters*, 10:47, 2010.

57. K.-H. Su, Q.-H. Wei, X. Zhang, J. J. Mock, D. R. Smith, and S. Schultz. Interparticle coupling effects on plasmon resonances of nanogold particles. *Nano Letters*, 3:1087, 2003.

58. Prashant K. Jain, Wenyu Huang, and Mostafa A. El-Sayed. On the universal scaling behavior of the distance decay of plasmon coupling in metal nanoparticle pairs: A plasmon ruler equation. *Nano Letters*, 7:2080, 2007.

59. Shengli Zou and George C. Schatz. Narrow plasmonic/photonic extinction and scattering line shapes for one and two dimensional silver nanoparticle arrays. *J. Chem. Phys.*, 121:12606, 2004.

60. Stefan A. Maier, Pieter G. Kik, and Harry A. Atwater. Observation of coupled plasmon-polariton modes in Au nanoparticle chain waveguides of different lengths: Estimation of waveguide loss. *App. Phys. Lett.*, 81:1714, 2002.

61. Mark L. Brongersma, John W. Hartman, and Harry A. Atwater. Electromagnetic energy transfer and switching in nanoparticle chain arrays below the diffraction limit. *Phys. Rev. B*, 62:R16356, 2000.

62. Stefan A. Maier, Pieter G. Kik, and Harry A. Atwater. Optical pulse propagation in metal nanoparticle chain waveguides. *Phys. Rev. B*, 67:205402, 2003.

Chapter 3
Experimental Methods

The subject of this thesis is the experimental study of 2-dimensional arrays of gold nanoparticles deposited on a self-organized insulating substrate. The fabrication and the optical/morphological characterization has represented a key point the research activity.

The fabrication of the systems has required the development of a suitable experimental setup dedicated to the growth and handling of metallic and insulating materials. The analysis has been performed by AFM for the morphological part, and various optical methods otherwise.

In this chapter we will first describe the molecular beam epitaxy (MBE) chamber assembled for the fabrication of the self-organized nanostructures, and then we will present a brief introduction to the experimental techniques employed for the morphological and optical characterizations of the specimens.

3.1 Preparation Chamber

The preparation and growth of the samples have been performed in a custom high vacuum (HV) vessel, designed and assembled in the course of the thesis for the specific requirements of this work. The chamber includes, in particular, equipment for the epitaxial growth of LiF and Au films and the thermal treatment of the specimens under in situ conditions.

The setup is schematically shown in Figs. 3.1 and 3.2. The HV chamber was optimized to enable a fast restoring of the base pressure after the frequent sample replacements; in fact, all the sample characterizations are performed ex situ, implying that the vacuum has to be broken relatively often. For this reason, we opted for a vessel with a relatively small frame, having a volume of $\approx 5\,L$, in order to reduce the pumped volume to the minimum. The pumping stage consists of a turbomolecular pump supported by a rotary pump for roughing. This allows a base pressure of $\approx 5 \times 10^{-8}$ mbar to be recovered within about 8 h after sealing, or a night span.

L. Anghinolfi, *Self-Organized Arrays of Gold Nanoparticles*, Springer Theses,
DOI: 10.1007/978-3-642-30496-5_3, © Springer-Verlag Berlin Heidelberg 2012

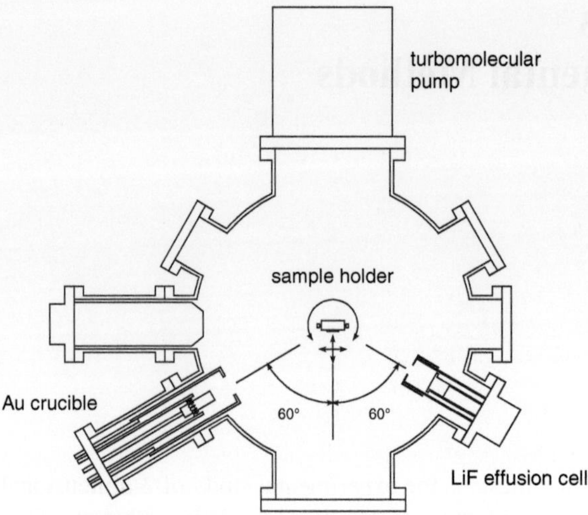

Fig. 3.1 Sketch of the experimental chamber (*front view*). The LiF effusion cell and the Au crucible are visible. The sample holder can rotate and translate in the plane of the figure for the fine positioning during the MBE depositions

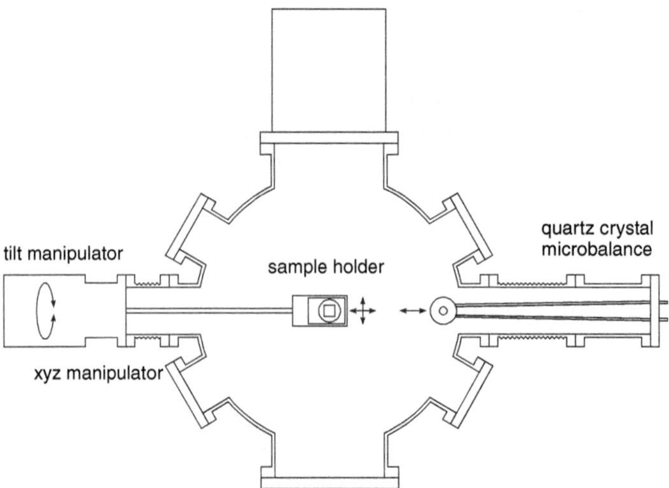

Fig. 3.2 Sketch of the experimental chamber (*side view*). The quartz crystal microbalance can be extended at the sample position, facing the LiF effusion cell or the Au crucible

The chamber is equipped with an ion gauge for pressure monitoring, a quartz crystal microbalance (QCM) for deposition rate calibrations, a home-made sample holder, an ion gun for sputtering, the Au and LiF MBE sources, along with several viewports to inspect the inside of the chamber, and perform in situ optical reflectance measurements.

The sample holder is supported on a 3-degrees of freedom xyz-translation stage, allowing the fine positioning inside the chamber with respect to the sources. It allows to anneal the sample up to $\approx 650\,°\text{C}$ maximum temperature.

The deposition of insulating or metallic films is performed by means of the *molecular beam epitaxy* (MBE) technique, which allows the formation of high quality crystalline films and nanostructures.

The LiF source is composed by a boron nitride (h-BN) crucible, in which LiF lump material to be deposited is loaded. h-BN is a ceramic material with the same layered structure of graphite, from which it is sometimes called white graphite; it has a lower electric conductance with respect to graphite, but its thermal and chemical stability are much superior. The crucible is supported by a W filament, that also serves to its heating by Joule effect. It reaches a temperature of $\approx 700\,°\text{C}$, sufficient to promote the sublimation of the lump LiF crystals (the melting point of LiF is at $845\,°\text{C}$). A screen of tantalum is used as collimator for the molecular beam and to prevent the chamber contamination and selectively expose the sample to the beam. The LiF source can achieve deposition rates up to at least 6 nm/min, at base pressure in the 10^{-7} mbar range. For this thesis, typical rates of 1 nm/min were chosen, as explained in more detail in the next chapter. The thickness of the deposited LiF layer was calibrated by ex situ optical characterization.

The Au source requires instead a different configuration, because of the higher Au melting point ($1,068\,°\text{C}$) and relatively low vapour pressure. In this case short sections of pure gold wires (99.99 %, MaTecK) are stored in a crucible of molybdenum, which is then heated by electron bombardment applying a high voltage with respect to a hot W filament. The crucible is supported by an alumina rod, which grants a good thermal insulation, and an additional copper screen is used both as collimator and thermal shield. The Au evaporator allowed to achieve evaporation rates in the 0.4–0.7 nm/min range, applying an overall power of $\approx 40\,\text{W}$ to the Mo crucible. Typical pressures during growth were in the 10^{-7} mbar range, that, though high, were perfectly acceptable due to the low reactivity of the involved material.

Although the characterization of the samples is typically performed mostly ex situ, it is possible to perform reflectivity measurements by means of an optical setup. The setup is sketched in Fig. 3.3: a white light beam, linearly polarized by a Glan-Thompson optical polarizer (usually s or p) is focused on the sample, at an incidence of $\theta = 60°$; after reflection on the sample, the beam is recollected by a second lens and analysed with a spectrometer (Ocean Optics USB2000+, wavelength range 400–1,000 nm). This configuration can be exploited for the real-time monitoring of the reflectivity during MBE growths or thermal treatments. A selected example of real-time monitoring of the sample's reflectivity will be shown in Sect. 5 (Fig. 5.8).

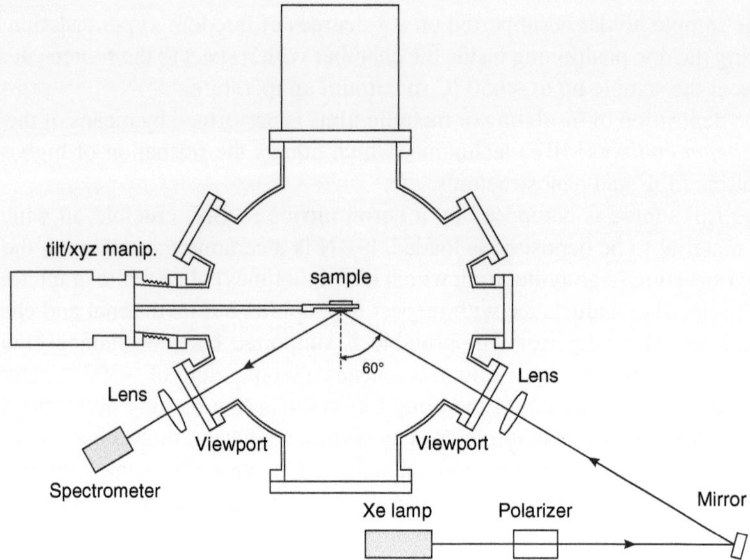

Fig. 3.3 Schematic diagram of the in situ reflectivity setup. See text for details

3.2 Optical Characterization

The optical characterizations of the samples fabricated during this thesis have been performed mainly ex-situ, by means of spectroscopic ellipsometry, reflectivity and transmission. In the next sections we will briefly review these techniques.

3.2.1 Spectroscopic Ellipsometry

Ellipsometry is a versatile and established technique, known since more than a century, primarily used for the evaluation of the optical constants and the thin-film thickness of samples. It is usually employed as a surface analysis tool, as it characterizes the reflection of light from surfaces, measuring the change of light polarization induced by the sample.

One of the remarkable features of spectroscopic ellipsometry is the high precision of the measurements [1–3], allowing, for example, sub-Angstroms thickness sensitivity in self-assembled monolayers [4–7]. It is also quite fast, usually requiring only a few seconds per scan, so even real-time observations can be easily carried out. Moreover, being an optical method, it is not destructive and can be performed in liquid environments.

As a drawback, ellipsometry data analysis follows an inherently indirect approach. Spectroscopic ellipsometry measurements consist of the determination of the so-called Ψ and Δ values (2.25) as a function of wavelength, and often at several angles of incidence; usually the measurements are carried out in the visible and near-UV region, but the range can also be extended to near-IR. As described in Sect. 2.1.2, Ψ and Δ represent the amplitude ratio and phase variations between the p and s components of the reflected light; by convention, they are written as

$$\rho = \tan \Psi \, e^{i\Delta} = \frac{r_p}{r_s} \tag{2.25}$$

where r_p and r_s are the complex Fresnel coefficients (2.23) . Separating the amplitudes and the phases, we find

$$\tan \Psi = \frac{|r_p|}{|r_s|} \tag{3.1a}$$

$$\Delta = \delta_p - \delta_s \tag{3.1b}$$

Looking at these definitions, we can expect that the direct interpretation of Ψ and Δ spectra is quite difficult. While certain optical features can be determined by a fast inspection of the spectra, the analysis typically requires the definition of an optical model, which describes the optical constants and layer thickness of the sample, and then is employed to reproduce the experimental data. In extreme cases, one has to construct an optical model even when the sample structure is not clear at all. Another disadvantage, which however in most cases is not limiting, is the low spatial resolution of the measurement, due to the finite spot size of the light beam, typically of few millimeters diameter.

Since the first null ellipsometers [1] working at a single wavelength, several spectroscopic configurations have been developed, allowing the fast acquisitions of spectra over a discrete range of frequencies; detailed reviews can be found in text books [1–3]. A general ellipsometer arrangement is sketched in Fig. 3.4. A well-collimated beam, coming from a white light source, reflects from the sample surface and is collected by a detector. Before reaching the sample, the beam passes through a polarizer, to produce controllable polarized light, and optionally through a compensator. The sample modifies the light polarization, and a second polarizer (called analyser) is placed before the detector to select the polarization state to analyse. Depending on the specific configuration, the polarizer, the analyser or the compensator can be rotating.

The ellipsometer used in this thesis is an M2000-U variable angle spectroscopic ellipsometer by J.A. Woollam Co., based on patented Diode Array Rotating Compensator Ellipsometer (DARCE) technology [8], and featuring the NIR module update. This instrument works in the wavelengths range between 245 and 1,680 nm, with a resolution of about 2 nm, and it also allows the acquisition of reflectance and transmittance spectra. As ellipsometer, it is based on a rotating compensator configuration, including a light source, a fixed polarizer, a rotating compensator, a sample holder stage, a fixed analyser and a detector stage (Fig. 3.4).

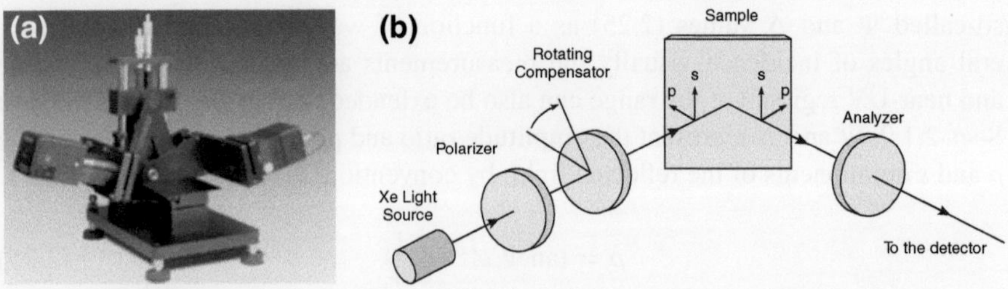

Fig. 3.4 Panel **a**: M2000-U spectroscopic ellipsometer picture. Panel **b**: schematic representation of the M2000-U configuration

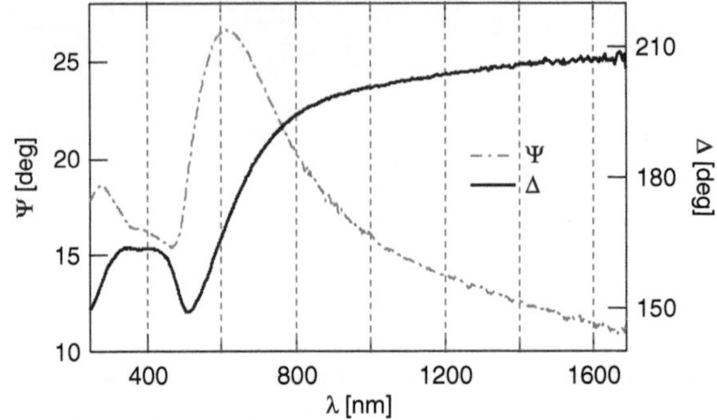

Fig. 3.5 Spectroscopic Ψ and Δ spectra, acquired at incidence of $\theta = 50°$ on a $\approx 6\,\mathrm{nm}$ film of gold supported on a nanopatterned LiF(110) substrate Fig. 3.10b

The instrument houses a 75 W Xe Arc lamp light source, emitting white unpolarized light in a broad continuous spectrum, from ultraviolet to near infrared. The polarizer and the analyser are two Glan-Taylor prisms [9]; for ellipsometric measurements they are fixed, while for reflectance and transmittance they can be rotated to adjust and select the polarization of the incident and reflected beams. The compensator is a complex multi-element pseudo-achromatic retarder [8]; the induced phase shift is not constant over the measured optical range, but has a slight dependence on the wavelength, so a rigorous calibration of retardation is provided to compensate for this error. Lastly, the detector consists of two spectrometers based on Diode Array technology; one of them covers the visible range up to $\approx 1{,}000\,\mathrm{nm}$, while the second one operates in the NIR region up to $\approx 1{,}700\,\mathrm{nm}$. Both the spectrometers include a dispersive optics stage, which spatially separates the incoming white light beam, and a photodiode array, which collects the diffracted light, with few nanometers of resolution; this allows the simultaneous acquisition of a total of 673 wavelengths.

An example of spectroscopic ellipsometry data is reported in Fig. 3.5, measured on a 6 nm Au film deposited on a nanopatterned LiF(110).

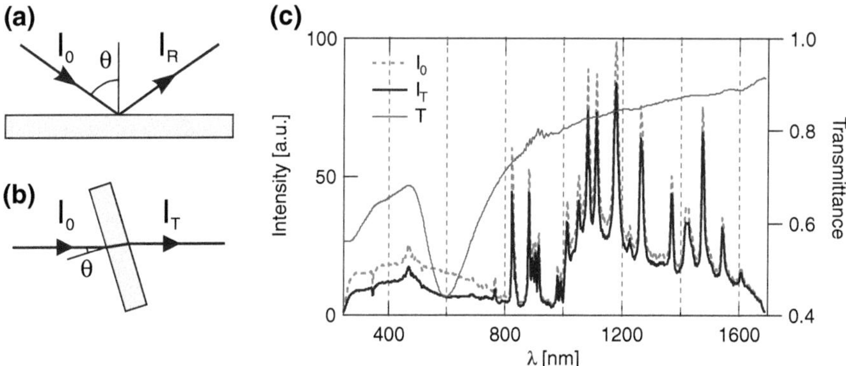

Fig. 3.6 Panels **a**, **b** sketches of reflectivity (**a**) and transmissivity (**b**) experiments, at incidence θ. Panel **c** example of baseline (*red curve*) and transmitted intensity (*black curve*) at normal incidence, and the corresponding transmittance obtained from their ratio (*blue curve*). The light source was the Xe lamp of the M2000-U ellipsometer; the sample was a $\approx$6 nm Au film deposited on a nanopatterned Lif(110) substrate Fig. 3.10b

3.2.2 Spectroscopic Reflectometry and Transmittance

We now briefly introduce the other optical methods employed in this thesis for the optical characterizations, namely reflectivity and transmissivity. Compared to ellipsometry, spectroscopic reflectivity (SR) and transmissivity (ST) are much simpler techniques, requiring a less elaborated experimental apparatus and allowing a more direct interpretation of the data; however, these measurements are also more affected by systematic errors and provide less informations on the optical properties of the samples.

The basic principle behind SR and ST experiments is quite straightforward. The intensity of a light beam is measured, as a function of the wavelength or energy, before (I_0, called *baseline*) and after the interaction with the sample under scrutiny; in case of SR, the intensity I_R of the reflected beam is measured Fig. 3.6a, while in case of ST the intensity I_T of the transmitted beam is acquired Fig. 3.6b. The ratios between the intensities of the reflected or transmitted beam and the incident beam are called the absolute reflectance R or transmittance T:

$$R(\omega) = \frac{I_R(\omega)}{I_0(\omega)}, \quad T(\omega) = \frac{I_T(\omega)}{I_0(\omega)} \tag{3.2}$$

An example of baseline and transmitted intensity, measured during a ST experiment, is reported in Fig. 3.6c, along with the computed transmittance T.

In general, a rudimentary SR/ST setup would require a source of light, a sample holder and a detector. In our case, the SR/ST measurements were performed employing the same M2000-U instrument used for spectroscopic ellipsometry, implying that the polarization of the light beam can be selected between s or p. First, the baseline

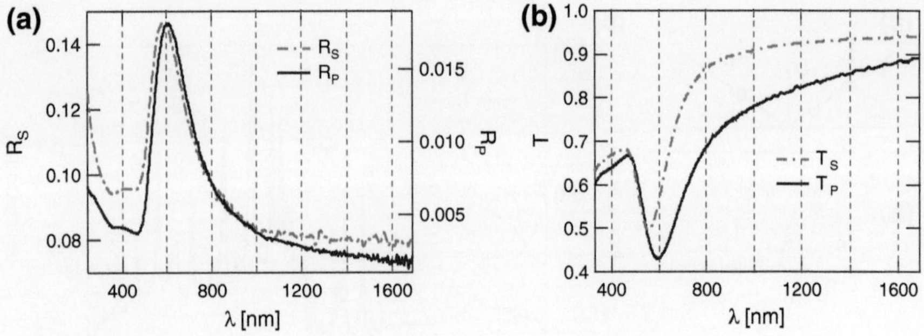

Fig. 3.7 Panel **a** s- and p-polarized light reflectivity, at $\theta = 50°$ of incidence, measured on a $\approx$6 nm Au film deposited on a nanopatterned Lif(110) substrate Fig. 3.10b. Panel **b** s- and p-polarized light transmissivity measured on the same sample at normal incidence

I_0 is acquired by measuring the beam emitted by the lamp directly on the detector, then, after inserting the sample, the reflected or transmitted beam is acquired, and R or T is computed according to (3.2) .

SR/ST measurements present several advantages and disadvantages with respect to spectroscopic ellipsometry. SR/ST require the acquisition of plain constant intensities, in contrast to ellipsometry where Ψ and Δ are computed from a time-dependent signal generated by the rotating element (polarizer, analyser or compensator); therefore, the experimental apparatus is much simpler, and custom setups can be easily implemented (see Fig. 3.3 for example). In addition, since the beam has a fixed s or p polarization, the measured reflectance and transmittance are related to only one reflection or transmission coefficient,[1] thus it is possible to separately investigate the effects of the two states of polarization. In contrast, the individual measurements provide less informations than ellipsometry, because the phases of the EM fields is lost by acquiring the intensity of light. Moreover, the baseline and the reflected/transmitted beams are acquired separately at different times, so the measurements are affected by the fluctuations in the lamp intensity, unlike in ellipsometry where the ratio $\rho = r_p/r_s$ is directly measured, so no baseline is required.

Some examples of SR/ST spectra are reported in Fig. 3.7, corresponding to the ellipsometric spectra in Fig. 3.5. Each curve in Fig. 3.7 represents a different measurement, in contrast to Fig. 3.5 where Ψ and Δ are acquired at the same time. We postpone the interpretation of the data to the next chapters. Here we just notice that from the SR/ST spectra a different optical behaviour of the system as a function of the polarization of light is clearly visible (two split sharp peaks for s- and p-polarized light), while in the ellipsometry spectra only one peak is present, broader and shifted in wavelength.

---

[1] I.e. $R_{s,p} = |r_{s,p}|^2$ and $T_{s,p} = |t_{s,p}|^2$, where $|\ldots|^2$ is the square magnitude of the complex value.

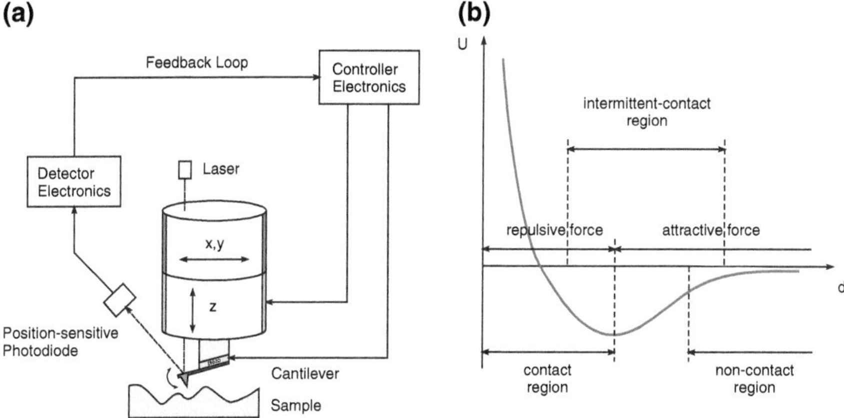

Fig. 3.8 Panel **a** sketch of a typical non-contact AFM instrument configuration. Panel **b** schematic representation of the probe-sample interaction potential and AFM operational modes

3.3 Atomic Force Microscopy

The morphological characterization of the samples has been performed by means of *atomic force microscopy* (AFM). This technique was developed for the first time in 1986 by Binnig et al. [10], and ever since has become a conventional and versatile method of surface analysis. AFM is a very high resolution type of scanning probe microscopy, capable of sub-A vertical resolution and lateral resolution down to atomic resolution. It is commonly employed for topographic imaging at nanoscale, but more advanced versions exists which also allow to probe, for example, surface potentials, electric conductance or magnetic domains.

A typical AFM setup is depicted in Fig. 3.8a. The microscope consists of a very sharp tip, with a few nanometers apex, located near the end of an elastic cantilever. By using piezoelectric stages, the tip is brought in close proximity to the sample, until an attractive or repulsive interaction is established. This force leads to a proportional deflection of the cantilever, which is detected by means of a laser spot reflected from the back of the cantilever onto a position-sensitive photodiode. The relative position between the tip and the sample is controlled by a feedback loop, which drives the piezoelectric scanners according to the deflection of the cantilever, and generates the actual AFM images by maintaining at constant level a particular operational parameter.

Depending on the specific situation, several kinds of forces can be measured. A typical probe-sample interaction potential is shown in Fig. 3.8b, as a function of the relative separation. When the tip is far from the sample, more than hundreds of nanometers, attractive long-range interactions prevail, like electrostatic or magnetic dipolar forces. Approaching the surface the attractive forces increase, and at a distance of the order of few nanometers the major contribution is from intermolecular Van

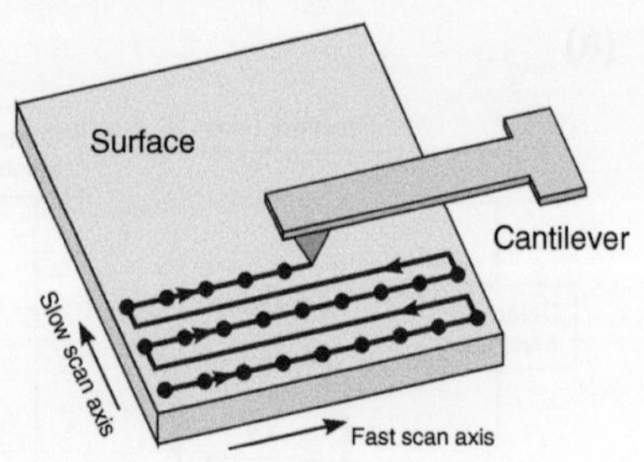

Fig. 3.9 Sketch of an AFM acquisition. The sample is scanned line-by-line in the fast axis direction and the tip-sample interaction is measured at a discrete set of points (*black circles*). At the end of each scan, the tip is shifted along the slow axis and brought back at the start of the line

der Waals interactions. Reducing further the distance, the tip starts to penetrate the sample and locally deforms the surface; elastic repulsion forces are now dominating, therefore the potential reaches a minimum, where the attractive and repulsive forces are equally balanced, and then rapidly diverges.

AFM acquisitions are performed by running the tip over a specified area of the sample, and measuring the interaction with the surface at a discrete set of points, as sketched in Fig. 3.9. Depending on the kind of measurement and on the nature of the sample, several operational modes are available. In order to acquire topographic images of surfaces the most common ones are *contact*, *non-contact* and *tapping* modes.

In *contact* AFM [10], the tip works in the repulsive range of the interaction potential and is dragged across the surface. During the scanning, the deflection of the cantilever is kept constant, and the corrections applied to the vertical piezoelectric stage provide the topographic information. This technique is usually fast and very accurate, but for soft or very rough samples the strong repulsive forces (of the order of nN) can lead to irreversible degradations of both the tip and the surface.

Non-contact mode [11] resolves this issue by increasing the tip-sample separation to about 50–150 Å, in the range of the attractive Van der Waals forces. These forces are considerably weaker than the ones established in contact mode, so the tip is given a small oscillation, and AC detection methods are used to detect the small changes in amplitude, phase, or frequency of the oscillating cantilever, in response to force gradients from the sample. This mode has the advantage that the tip is never in contact with the sample and therefore can not sketch or destroy it, however it provides lower resolution and in presence of a fluid contaminant layer the oscillating probe can be trapped in the fluid, failing to image the true surface of the sample.

Tapping AFM is a high resolution technique which overcomes both the problems associated with the degradation of the sample and with friction and adhesion. Like non-contact AFM, the tip is not static, however the oscillation amplitude is much higher (typically in the range 20–200 nm), the tip being alternately in contact with the surface, to provide high resolution, and lifted off, to avoid the alteration of the surface. As the tip approaches the sample, the force gradient alters the cantilever motion, modifying the amplitude, the resonance frequency, and the phase of the oscillations. During scanning, the feedback system adjusts the relative distance between the tip and the sample in order to maintain one of these parameters at a constant level,

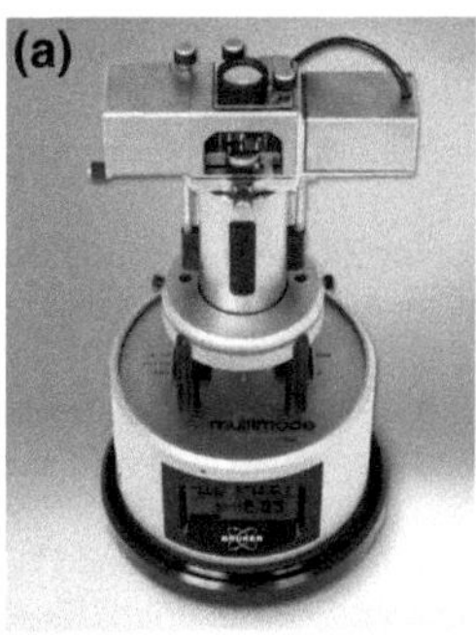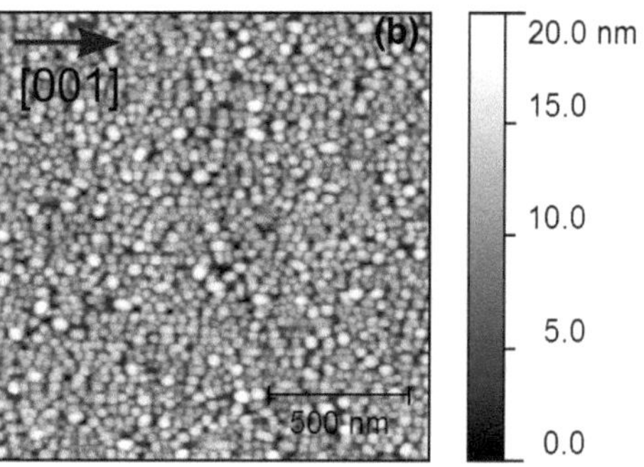

Fig. 3.10 Panel **a** picture of a multimode scanning probe microscope. Panel **b** AFM image of a 6 nm Au film on a nanopatterned LiF(110) acquired in tapping mode

and correspondingly generates the AFM image. Usually the topographic surface features are detected by monitoring the oscillation amplitude; instead, by inspecting the oscillation phase variations it can be achieved a better contrast on material surface properties, such as stiffness, viscoelasticity, and chemical composition [12–16].

The AFM setup employed in this work was composed of a Multimode Scanning Probe Microscope (Fig. 3.10a, maximum scan size 15 µm) driven by a Nanoscope IV controller, both produced by Digital Instruments-Veeco, and all the images were acquired in tapping mode; an example is shown in Fig. 3.10b.

References

1. H. Fujiwara. *Spectroscopic Ellipsometry. Principles and Applications*. Wiley, 2007.
2. Bashara N.M. Azzam R.M.A. *Ellipsometry and Polarized Light*. North Holland, 1988.
3. H. Tompkins and E. Irene. *Handbook of Ellipsometry*. Noyes Publications, 2005.
4. Mirko Prato, Riccardo Moroni, Francesco Bisio, Ranieri Rolandi, Lorenzo Mattera, Ornella Cavalleri, and Maurizio Canepa. Optical characterization of thiolate self-assembled monolayers on Au(111). *J. Phys. Chem. C*, 112:3899, 2008.
5. F. Bordi, M. Prato, O. Cavalleri, C. Cametti, M. Canepa, and A. Gliozzi. Azurin self-assembled monolayers characterized by coupling electrical impedance spectroscopy and spectroscopic ellipsometry. *J. Phys. Chem. B*, 108:20263, 2004.
6. Mirko Prato, Marina Alloisio, Sushilkumar A. Jadhav, Andrea Chincarini, Tiziana Svaldo-Lanero, Francesco Bisio, Ornella Cavalleri, and Maurizio Canepa. Optical properties of disulfide-functionalized diacetylene self-assembled monolayers on gold: a spectroscopic ellipsometry study. *J. Phys. Chem. C*, 113:20683, 2009.
7. Hicham Hamoudi, Zhiang Guo, Mirko Prato, Celine Dablemont, Wan Quan Zheng, Bernard Bourguignon, Maurizio Canepa, and Vladimir A. Esaulov. On the self assembly of short chain alkanedithiols. *Phys. Chem. Chem. Phys.*, 10:6836, 2008.
8. Daniel W. Thompson Blaine D. Johs. Regression calibrated spectroscopic rotating compensator ellipsometer system with photo array detector. Patent, 1999.
9. J F Archard and A M Taylor. Improved Glan-Foucault prism. *J. Sci. Instr.*, 25:407, 1948.
10. G. Binnig, C. F. Quate, and Ch. Gerber. Atomic force microscope. *Phys. Rev. Lett.*, 56:930, 1986.

11. Y. Martin, C.C. Williams, and H.K. Wickramasinghe. Atomic force microscope force mapping and profiling on a sub 100-ascale. *J. Appl. Phys.*, 61:4723, 1987.
12. S.N. Magonov, V. Elings, and M.-H. Whangbo. Phase imaging and stiffness in tapping-mode atomic force microscopy. *Surf. Sci.*, 375:L385, 1997.
13. William W. Scott and Bharat Bhushan. Use of phase imaging in atomic force microscopy for measurement of viscoelastic contrast in polymer nanocomposites and molecularly thick lubricant films. *Ultramicroscopy*, 97:151, 2003.
14. Tisato Kajiyama, Keiji Tanaka, Isao Ohki, Shou-Ren Ge, Jeong-Sik Yoon, and Atsushi Takahara. Imaging of dynamic viscoelastic properties of a phase-separated polymer surface by forced oscillation atomic force microscopy. *Macromolecules*, 27(26):7932, 1994.
15. C. Daniel Frisbie, Lawrence F. Rozsnyai, Aleksandr Noy, Mark S. Wrighton, and Charles M. Lieber. Functional group imaging by chemical force microscopy. *Science*, 265:2071, 1994.
16. Aleksandr Noy, Dmitri V. Vezenov, and Charles M. Lieber. Chemical force microscopy. *Ann. Rev. Mat. Sci.*, 27:381, 1997.

Chapter 4
Self-Organized Nanoparticle Arrays: Morphological Aspects

In this chapter we present the procedures for the fabrication of lithium fluoride nanostructured substrates and for the subsequent realization of ordered 2D arrays of metallic nanoparticles.

4.1 Growth and Characterization of Lithium Fluoride Substrates

Wide band gap insulators have been extensively employed as templates for the guided overgrowth of metallic structures since a long time [1–6]. Insulating surfaces provide a complete electronic decoupling, thus metallic adsorbates can develop a local electronic structure with specific functionality depending on the growth conditions. Moreover, wide band gap solids are transparent from the near-UV to the near-IR range, so they easily allow the investigation of the optical properties of the adsorbates, are quite stable in atmosphere, do not suffer oxidation (but in some cases are hygroscopic), and are usually non magnetic.

Ionic crystals, such as NaCl, LiF, MgO, CaF_2, KBr, are a class of insulators composed of alternating ions of opposite sign. The ions are bound together mainly by electrostatic attractions, so the electrostatic energy strongly affects the stability of the surfaces, making certain crystal orientations highly favoured. This peculiarity is commonly exploited for the patterning of surfaces at nanoscale [1–6]: when the crystal is cut along an unfavourable direction and the atoms are given enough mobility, the less stable surfaces can undergo faceting in favour of the more stable orientations, spontaneously leading to the formation of regular structures. For example, faceting into ridges or pyramids, or the formation of regular ridge-and-valley structures are common phenomena observed upon thermal annealing or homoepitaxy on NaCl(111) [6], NaCl(110) [7, 8], LiF(110) [5], CaF_2(110) [5] or MgO(110) [9].

In this thesis, we investigated and characterized the homoepitaxial growth on LiF(110). LiF crystals are composed of Li^+ and F^- ions arranged on a rocksalt

L. Anghinolfi, *Self-Organized Arrays of Gold Nanoparticles*, Springer Theses, DOI: 10.1007/978-3-642-30496-5_4, © Springer-Verlag Berlin Heidelberg 2012

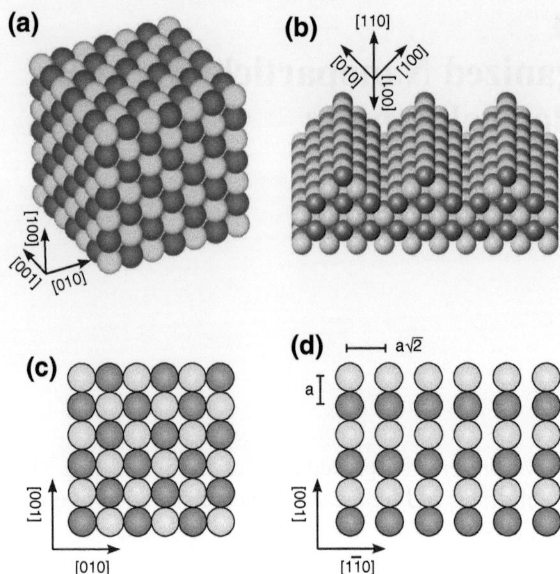

Fig. 4.1 Panel **a** 3-dimensional view of a cubic crystal of lithium fluoride; atoms of lithium and fluorine denoted with different colours. Panel **b** 3-dimensional representation of a portion of ripple structure developed on LiF(110). Panel **c** top view of the LiF(100) surface. Panel **d** top view of the LiF(110) surface

cubic structure, with each ion surrounded by six ions of the other species Fig. (4.1). The preferential orientations of the crystals are the {100} surfaces, consisting of alternating ions arranged on a square lattice (Fig. 4.1c). {110} surfaces are instead less symmetric and less stable, the ions forming parallel chains with alternating charge signs Fig. 4.1d . Therefore, LiF crystals cut along the {110} direction expose unstable surfaces, and during homoepitaxial deposition of LiF they can undergo a spontaneous surface reconstruction driven by the proliferation of {100} facets; these facets are tilted of 45 ° with respect to the surface, so that elongated islands with [001] macrosteps are formed, which eventually develop into a regularly spaced ridge-and-valley (*ripple*) structure Fig. 4.1b [5].

Mechanically-polished optical-grade LiF(110) substrates, with lateral sizes of 1 cm and 1 mm thick, were acquired from MaTecK Gmbh. Some samples were manufactured with both sides polished and were used for transmission measurements; the remaining ones, more suited to reflectivity measurements, were polished on only one side, the other one being rough in order to minimize the reflection from the backside.

A representative AFM image of a LiF(110) substrate in the as-received state is reported in Fig. 4.2. As said before the (110) surface is not a preferred crystal plane, so it does not present extended flat terraces, but instead is uniformly rough. The straight scratches visible in the figure are due to the polishing procedure and were

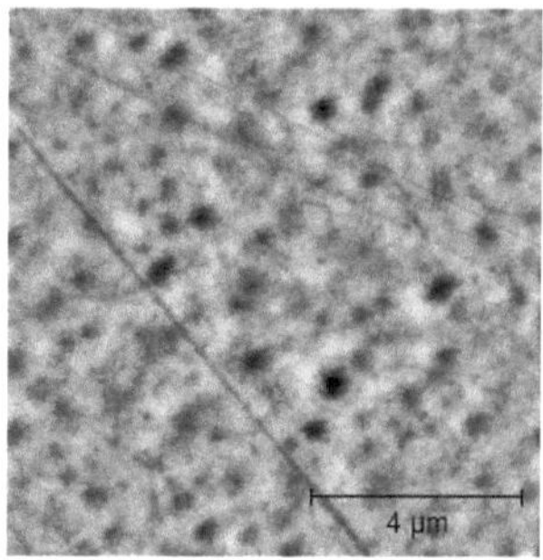

Fig. 4.2 Representative AFM image of a LiF(110) (MaTecK Gmbh) substrate in the as-received state

observed on all the samples; in any case, they did not compromise the fabrication of the templates. A statistical analysis of the surface characteristics, performed over several samples, yielded a RMS surface roughness of about 6–9 nm. These values were adequate to induce the formation of regular structures upon LiF homoepitaxy; in contrast, samples exhibiting RMS roughness values higher than 15 nm were found unsuitable for the fabrication of ordered nanostructures.

After the initial characterization, the specimens were loaded in the HV chamber. Prior to the deposition, they were annealed in vacuum at 450 °C for about 1 h, in order to remove any surface contamination, and then cooled down to the desired growth temperature. The substrates did not exhibit any faceting following the annealing. The homoepitaxial growth was performed at normal incidence, as sketched in Fig. 4.3a, under high vacuum conditions (pressures lower than 5×10^{-7} mbar), and placing the effusion cell at approximately 10 cm from the sample surface.

The deposition rate was a critical parameter for the achievement of well-ordered nanostructures. In Fig. 4.3b, c we report AFM images measured ex situ on two different samples after the deposition of LiF at a rate of 4 nm/min and 1 nm/min, respectively; in both cases the substrate temperature was kept constant at $\approx$300 °C and approximately 250 nm of LiF were deposited. For the sample in panel (c) (1 nm/min) an ordered ripple morphology is clearly visible, showing parallel and coherently aligned [001] macrosteps with an average spacing of $\approx$35 nm, and facets at an angle of 45 ° with respect to the sample plane (panel (c)). The morphology of the sample in panel (b) (4 nm/min), instead, is dramatically different. Here the surface is completely chaotic, consisting of randomly oriented rectangular prisms with {100} facets: in this case the deposition rate was so high to prevent either long range ordering but also any local crystallographic matching with the substrate, leading to the proliferation of stand-alone nanosized crystals. Therefore, we fixed the deposition rate at 1 nm/min for all the samples under scrutiny in this work.

In order to investigate the evolution of the surface morphology from flat to nanostructured, we monitored the surface morphology at different stages of deposition. AFM images measured in correspondence of increasing deposition times are shown in Fig. 4.4, at thickness step of $\approx$60 nm. The [100] facets are already visible at the lowest coverage of 60 nm (panel a), forming elongated structures a few tens of nanometers long and oriented along the [001] direction; at this stage, however, the macrosteps are

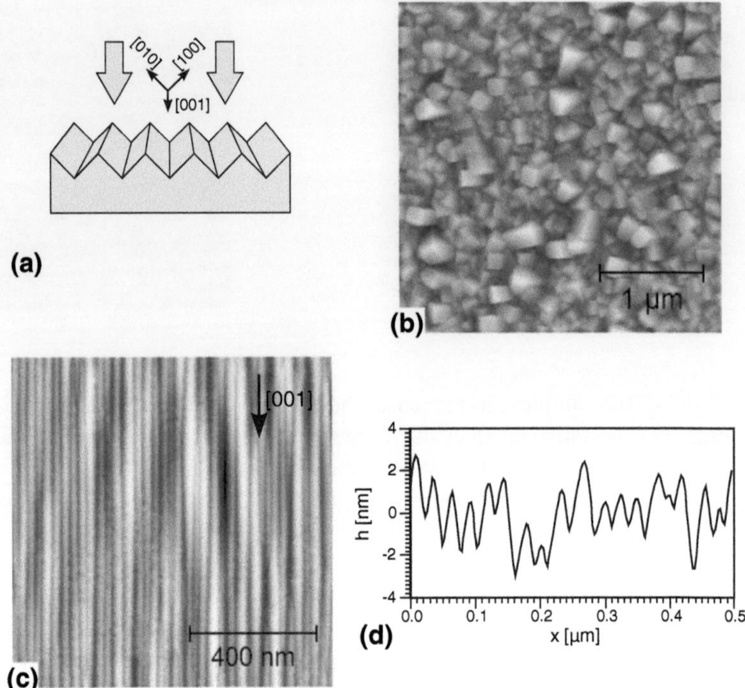

Fig. 4.3 Panel **a** sketch of the LiF deposition at normal incidence on LiF(110), and the subsequent formation of LiF ripples. Panel **b** AFM image of a LiF(110) sample after the deposition of LiF at $T = 300\,°C$ and rate of $4\,nm/min$. Panel **c** AFM image of a nanopatterned LiF(110) sample with periodicity $\Lambda = 35\,nm$, after the deposition of $t_{LiF} = 250\,nm$ LiF at a rate of $1\,nm/min$ and $T = 300\,°C$. Panel **d** representative AFM profile of the image in panel **c**

still relatively randomly distributed and no univocal long range order can be distinguished. The corresponding Fourier spectral density is shown in the inset of Fig. 4.4a: the spectrum is predominantly elongated along the $[1\bar{1}0]$ direction, indicating features of the real image with preferential [001] orientation, but it is also spread in the [001] direction, confirming the lack of long range order. Increasing the coverage, as in panels (b) and (c), the macrosteps get longer, merging with one another and eventually developing into a regular ripple structure; correspondingly, the Fourier spectra become progressively sharper and confined in the $[1\bar{1}0]$ direction. After the deposition of $\approx 250\,nm$ of LiF (panel d), the ripple periodicity is well defined and the [001] macrosteps are few μm long. At even higher coverages, the quality of the ripples does not change significantly, but several defects begin to proliferate, mainly in proximity of the polishing scratches, leading to the formation of randomly oriented nanocubes, similar to Fig. 4.3b.

The macrosteps periodicity Λ and, in general, the structure of the ripples are determined by the adatoms mobility and diffusion length during the homoepitaxy, so they strongly depend on the substrate temperature and the deposition rate. Having

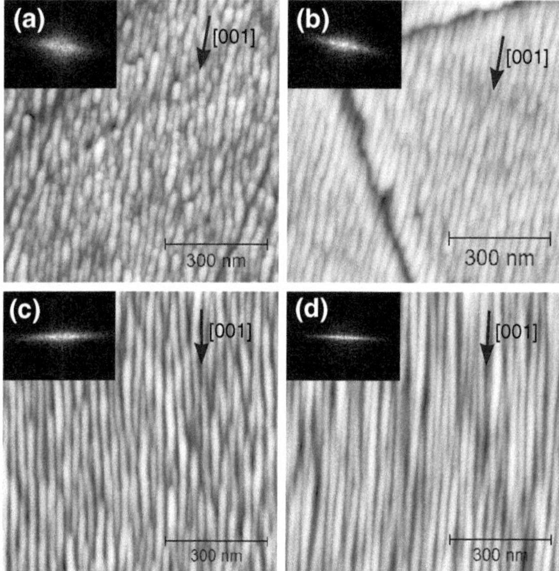

Fig. 4.4   AFM images of a nanopatterned LiF(110) sample with $\Lambda = 35$ nm, as a function of the deposited amount of LiF. Panel **a** $t_{LiF} = 60$ nm. Panel **b** $t_{LiF} = 120$ nm. Panel **c** $t_{LiF} = 180$ nm. Panel **d** $t_{LiF} = 250$ nm. Inset of the figures: Fourier spectra of the corresponding images.

fixed the rate at ≈ 1 nm/min, we focused on the study of the periodicity as a function of the temperature T. In Fig. 4.5 we report various AFM images measured on several samples after the deposition of ≈ 250 nm of LiF at different temperatures, within the range between 250 and 400 °C. In all cases the ripple structure is clearly recognizable and quite regular. The ripples are very elongated especially at the lowest temperatures ($T = 250$ °C for panel a and $T = 300$ °C for panel b), where they extend longer than the image size (1 µm) and are very straight. This is confirmed by the corresponding 2D Fourier spectra, which rapidly decay in the [001] direction and are much broader in the [1$\bar{1}$0]. At higher temperatures, $T = 350$ °C for panel (c) and $T = 400$ °C for panel (d), the ripples are more widely spaced but also more irregular; the average length is in-between 500 nm and 1 µm, and the grooves appear slightly distorted. Looking at the 2D FFT spectra, the central peaks are still mainly spread in the [1$\bar{1}$0] direction, confirming that the uniaxial shape of the ripple remains the dominant feature of the surfaces, however an increasing broadening in the [001] direction is also observed, indicating a gradual degradation of the ripples with the temperature. The broadening of the 2D Fourier spectra in the [1$\bar{1}$0] direction decreases with the ripples periodicity, however this is not an effect of disorder; in fact, the Fourier spectrum intensity for an ideal sawtooth profile with period Λ decays proportionally to $1/\Lambda^2$ [10].

From a statistical analysis of the AFM data, we could extract the average ripple periodicity Λ along with its standard deviation; we report the results as a function

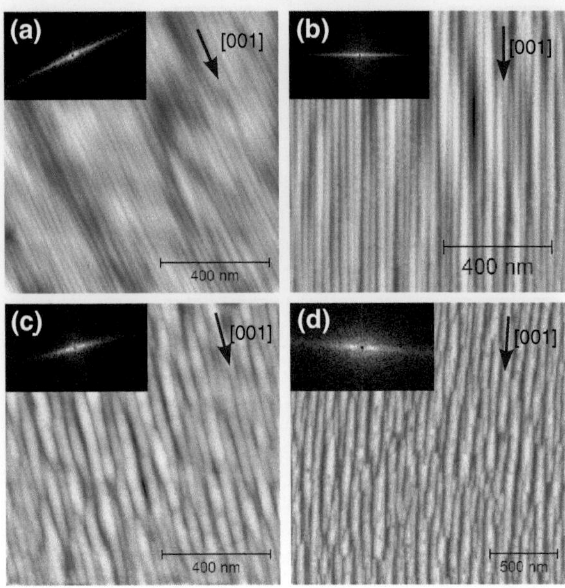

Fig. 4.5 AFM images of nanopatterned LiF(110) samples after the deposition of $t_{LiF} \approx 250$ nm of LiF, grown at different temperatures. Panel **a** $T = 250\,°C$, $\Lambda = (25 \pm 2)$ nm. Panel **b** $T = 300\,°C$, $\Lambda = (35 \pm 3)$ nm. Panel **c** $T = 350\,°C$, $\Lambda = (45 \pm 5)$ nm. Panel **d** $T = 400\,°C$, $\Lambda = (60 \pm 7)$ nm. Inset of the figures: Fourier spectra of the corresponding images

Fig. 4.6 Ripples periodicity as a function of the substrate temperature during the LiF homoepitaxy. *Dashed line* linear trend of ≈ 1 nm/5 °C to guide the eye

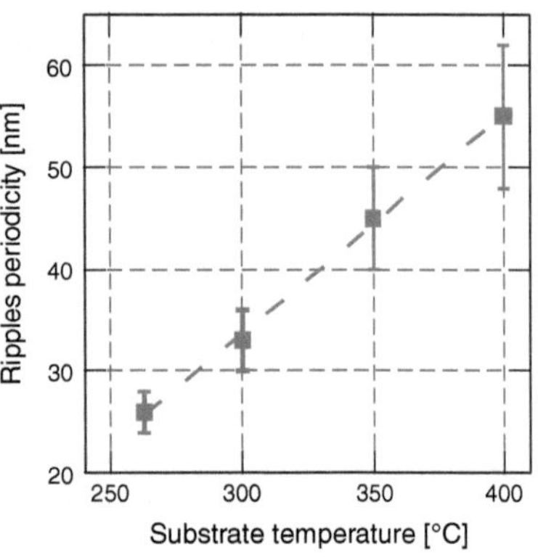

of T in Fig. 4.6. Λ follows an almost linear trend with the temperature, with a rate of approximately 1 nm/5 °C: at 250 °C the average ripple spacing is (25 ± 2) nm Fig. 4.5a; increasing T, Λ raises to (35 ± 3) nm at 300 °C Fig. (4.5b), (45 ± 5) nm at 350 °C Fig. (4.5c) and (60 ± 7) nm at 400 °C Fig. (4.5d); correspondingly, the standard deviation remains as low as 3 nm below 300 °C and increases up to ≈ 7 nm at a temperature of 400 °C.

We did not consider depositions at temperatures outside the 250–400 °C range. However, lowering the temperature the adatoms mobility decreases and so the faceting of the (110) surfaces is harder to occur; on the contrary, at high temperature we are on one side limited by the sample holder capabilities, on the other side the high adatoms thermal energy is expected to gradually destroy the ripple coherence.

4.2 2D Arrays of Gold Nanoparticles

We now address the fabrication of bidimensional arrays of metallic nanoparticles (NP), employing the nanostructured LiF substrates as templates to control their size and assist their arrangement. The fabrication of the NP arrays follows a two steps procedure. First a thin layer of gold is deposited on the rippled LiF templates, and then the NP array formation is induced by thermal annealing.

As sketched in Fig. 4.7a, the deposition of gold is performed at grazing incidence, in order to exploit the shadowing effect of the ripples ridges: if the molecular beam of gold reaches the LiF surface at an angle larger than $45\,°$ with respect to the surface normal and perpendicularly to the [001] direction, only one side of the grooves gets exposed to the beam, and an array of disconnected "wires" is formed. Similarly to the previous case, the deposition is carried out in high vacuum conditions, in order to limit any possible contamination of the metallic film. The molecular beam of gold is released from the Mo crucible with a rate of ≈ 0.5 nm/min, and impinges the sample at an angle of incidence of $60\,°$. In order to avoid the diffusion of the adatoms across the ripples ridges, the temperature is kept constant at approximately $100\,°$C and after the deposition the samples are immediately cooled down to room temperature. A maximum of ≈ 10 nm of gold has been deposited, as at higher coverages the ripples grooves get completely filled with gold and a continuous layer is formed.

In Fig. 4.7b we report an AFM image measured ex situ on the same sample of Fig. 4.5(c) after the deposition of ≈ 5 nm of gold. The surface is still characterized by coherently aligned elongated structures, indicating that the overgrown layer has followed the structure of the underlying template. The ordering is also confirmed by the 2D Fourier spectrum, where the central peak is still quite stretched in the $[0\bar{1}1]$ direction; however, a secondary background is also present, slowly decaying from the central peak in all directions uniformly. This indicates that the metallic stripes are not completely continuous but rather composed of close-packed grains; it also suggests that the growth of gold on LiF does not likely proceed layer-by-layer but probably follows a layer-plus-island (Stranski–Krastanov) mode. Due to the finite size of the AFM tip, it is not possible to deduce from the AFM images whether the islands coalesced or remained separated. Nevertheless, the lateral size of the grains seems to be limited to an upper size of ≈ 15 nm. This is shown in Fig. 4.7c, d, where we report AFM images measured on two samples with different ripple spacing and the same amount of deposited gold, ≈ 5 nm; in panel (c), the sample has a periodicity $\Lambda \approx 25$ nm and linear chains of grains are observed, where the grains lateral sizes

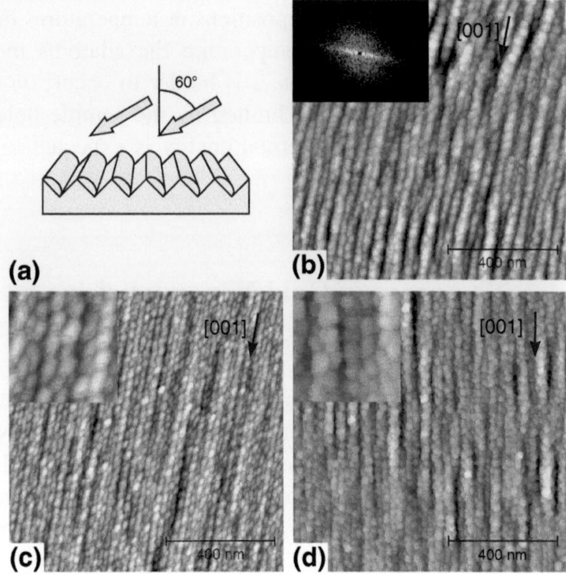

Fig. 4.7 Panel **a** sketch of the grazing deposition of Au on rippled LiF(110). Panels **b**, **c**, **d** AFM images of nanopatterned LiF(110), with different periodicities, after the deposition of $t_{Au} \approx 5$ nm of gold at $T \approx 100\,°C$; $\Lambda = 45$ nm (**b**, Fourier spectrum in the inset), $\Lambda = 25$ nm (**c**), $\Lambda = 40$ nm (**d**)

matches the grooves width; on the contrary, in panels (d) and (b), for a spacing $\Lambda > 40$ nm, no island as large as the ripples is found, and instead multiple chains are formed inside each groove.

Given the natural tendency of gold to form agglomerates, we can further promote the dewetting of the "nanowires" (NW) by mild annealing the substrates [11–13]. In Fig. 4.8c we report an AFM image measured on the same sample of Fig. 4.7b after the annealing at $400\,°C$. Comparing the two images, the main effect of the annealing procedure was a partial melting of the gold grains, which merged together forming larger particles, with a size comparable to the ripples width. The underlying ripple structure assisted the process, preventing the spreading of the grains across the ridges, and instead promoting the formation of parallel chains of nanoparticles, preserving the same periodicity Λ of the ripples.

Interestingly, the LiF grooves not only guided the particles arrangement, but also influenced their sizes and shape. This effect can be observed looking at Fig. 4.8, where we reported AFM images of several NP arrays, annealed at the same temperature of $400\,°C$ but fabricated under different growth conditions. Let's first consider panels (a) and (b). The samples have, respectively, periodicities $\Lambda \approx 30$ nm and ≈ 40 nm, and a similar amount of gold were deposited, ≈ 4.5 and ≈ 4.8 nm. In both cases the particles have an in-plane circular shape, and, although it is not possible to precisely determine the particles dimensions due to the convolution of the AFM tip, their typical size increases according to the variations of width of the LiF grooves. In panel (c) a sample with $\Lambda \approx 45$ nm, similar to the sample in panel (b), but $t_{Au} \approx 5.4$ nm of deposited gold is shown. In this case, the particles are still confined inside the ripples, and the higher amount of gold led to the formation of more elongated structures, with

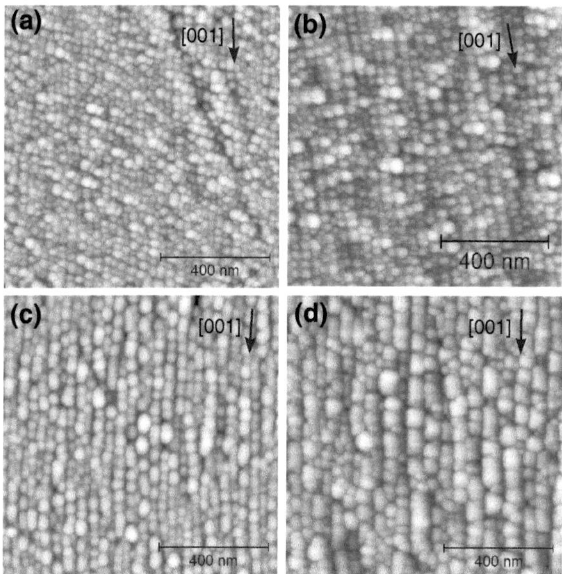

Fig. 4.8 AFM images of gold nanoparticles arrays on nanopatterned LiF(110), following the deposition of t_{Au} gold and annealing at $T \approx 100\,°C$. Panel **a** $\Lambda = 30\,$nm, $t_{Au} = 4.5\,$nm. Panel **b** $\Lambda = 40\,$nm, $t_{Au} = 4.8\,$nm. Panel **c** $\Lambda = 45\,$nm, $t_{Au} = 5.4\,$nm. Panel **d** $\Lambda = 55\,$nm, $t_{Au} = 7.2\,$nm

an average aspect ratio of ≈ 1.5. A similar trend is observed increasing further the thickness of gold. The sample in panel (d) has $\Lambda \approx 55\,$nm and $t_{Au} \approx 7.2\,$nm.

The majority of the particles is again as wide as the grooves, however we also observe an increased dispersion of particles sizes with respect to the previous cases, both along and across the ripples. This is probably due to the presence of morphological defects in the underlying ripple structure, and it happens in concordance with the observations of the previous section, where an increasing disorder of the surface morphology was found at the largest Λ Fig. (4.5d). Looking at the shape, the particles now have a maximum aspect ratio of ≈ 2, higher than for sample (c), but many less anisotropic ones are also visible; this could be due to the fact that the formation of structures with aspect ratios greater than two is not favoured, so in these cases a segregation to smaller particles occurs. Then, we can conclude that, for periodicities between 25 and 60 nm and thickness of deposited gold between 4 and 10 nm, the particles width is fixed by the grooves width while the shape (aspect ratio) is determined by the amount of deposited gold.

We can interpret this effects by considering the dynamics of the Au nanowires dewetting. If the NW evolve following a Rayleigh-type instability [14], then a linear relationship between the NP spacing (and hence their volume) and the initial NW radius r ($r \propto (\Lambda \cdot t_{Au})^{1/2}$) is expected [15], showing that both mean spacings along the ripples and volume of the NP get larger in correspondence of the increasing Λ and t_{Au}. The simultaneous increase in the aspect ratio can be ascribed to the lateral

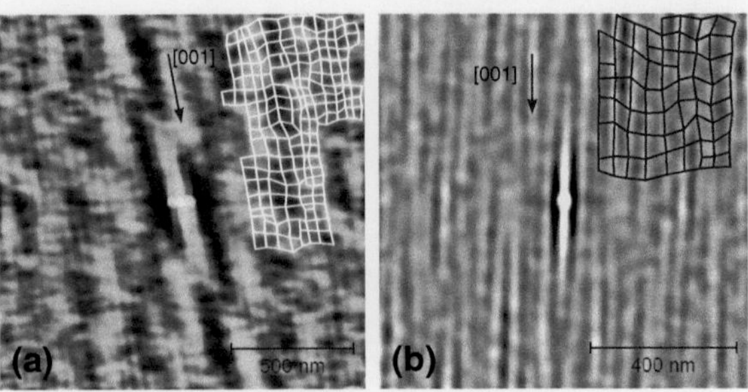

Fig. 4.9 2-dimensional autocorrelations of the AFM images in Fig. 4.8b and c. The superimposed grids are guide to the eye for the positions of the maxima of the correlation

"constraint" effect exerted by the LiF nanopatterns, that triggers an anisotropic NP growth during the dewetting process.

Another unexpected feature of the NP arrays is the relative positions of particles belonging to different grooves. As the dewetting induced by the annealing takes place separately in each ripple, one may expect no transversal correlation between the position of the particles in adjacent rows; in contrast, many regions can be distinguished where the particles are locally arranged on a rectangular grid. These regions can be emphasized by looking at the autocorrelations of the AFM images. In Fig. 4.9 we report the 2D autocorrelation plots for the samples of Fig. 4.8b, c; both figures are characterized by rectangular patterns of peaks and hollows, with sides oriented in the $[1\bar{1}0]$ and $[001]$ directions; the typical spacing along the $[1\bar{1}0]$ direction matches the ripples periodicity, while the separations in the $[001]$ direction are on the same scale but more irregular. These structures indicate that the NP arrays have a good translational symmetry not only in the direction across the ripples, as imposed by the regular spacing of the grooves, but also along the ripples, which is unexpectedly triggered by the annealing. The extension of the rectangular NP meshes depends on the dispersion of the particles size; it can be quite large in case of monodispersed samples like in Fig. 4.8a, b, where the majority of the visible particles are orderly arranged, while for samples like in Fig. 4.8d almost no inter-ridge correlation is found.

Lastly, for less than ≈ 3 nm of deposited gold a different dewetting mode occurs. Due to the low amount of gold, the particles cannot grow as wide as the ripples, and therefore they form multiple chains inside each groove, similar to the situation observed before the annealing Fig. 4.7b, d. We report an example in Fig. 4.10; the sample has a periodicity $\Lambda \approx 30$ nm and ≈ 2.7 nm of gold were deposited. We used dashed lines to delimit the grooves in the figure, so that double or triple chains of particles can be distinguished. Furthermore, looking at the 2D autocorrelation plot in panel (b) we can identify again a weak rectangular pattern from the NP array, indicating that even in this case the thermal dewetting favoured a partial alignment of the particles.

Summarizing, we have observed that it is possible to fabricate regular arrays of noble-metal NPs on insulating substrates by means of self-organization approaches,

Fig. 4.10 Panel **a** AFM image of a nanopatterned LiF(110) sample after the deposition of $t_{Au} = 2.7$ nm Au at $T \approx 100\,°C$ and annealing at $T = 400\,°C$; the *vertical dashed lines* emphasizes the periodicity of the ripples. Panel **b** 2D autocorrelation of the image in panel **a**

in which important morphological parameters, like the mean NP size, shape and mutual spacing, can be controlled by means of various fabrication parameters. This flexibility in NP array fabrication will be exploited to correspondingly tune their plasmonic response, as detailed in the following chapter.

References

1. R. Bennewitz. Structured surfaces of wide band gap insulators as templates for overgrowth of adsorbates. *J. Phys. Condens. Mat.*, 18:R417, 2006.
2. Jason R. Heffelfinger and C. Barry Carter. Mechanisms of surface faceting and coarsening. *Surf. Sci.*, 389:188, 1997.
3. L Nony, R Bennewitz, O Pfeiffer, E Gnecco, A Baratoff, E Meyer, T Eguchi, A Gourdon, and C Joachim. Cu-TBPP and PTCDA molecules on insulating surfaces studied by ultra-high-vacuum non-contact AFM. *Nanotechnology*, 15:S91, 2004.
4. R. C. Barrett and C. F. Quate. Imaging polished sapphire with atomic force microscopy. *J. Vac. Sci. Tech. A*, 8:400, 1990.
5. A. Sugawara and K. Mae. Surface morphology of epitaxial LiF(110) and CaF_2(110) layers. *J. Vac. Sci. Tech. B*, 23:443, 2005.
6. Akira Sugawara, G. G. Hembree, and M. R. Scheinfein. Self-organized mesoscopic magnetic structures. *J. App. Phys.*, 82:5662, 1997.
7. Akira Sugawara and K. Mae. Nanoscale faceting of a NaCl(110) homoepitaxial layer. *Journal of Crystal Growth*, 237–239:201, 2002.
8. Akira Sugawara, T. Coyle, G. G. Hembree, and M. R. Scheinfein. Self-organized Fe nanowire arrays prepared by shadow deposition on NaCl(110) templates. *App. Phys. Lett.*, 70:1043, 1997.
9. Akira Sugawara and K. Mae. Faceting of homoepitaxial MgO(110) layers prepared by electron beam evaporation. *Surf. Sci.*, 558:211, 2004.
10. K. Howell. *The Transforms and Applications Handbook*, chapter 2. CRC-Press, 2000.
11. Claudia Manuela Mller, Flavio Carlo Filippo Mornaghini, and Ralph Spolenak. Ordered arrays of faceted gold nanoparticles obtained by dewetting and nanosphere lithography. *Nanotechnology*, 19:485306, 2008.
12. N. Rougemaille and A. K. Schmid. Self-organization and magnetic domain microstructure of Fe nanowire arrays. *J. App. Phys.*, 99:08S502, 2006.

13. Yong-Jun Oh, Caroline A. Ross, Yeon Sik Jung, Yang Wang, and Carl V. Thompson. Cobalt nanoparticle arrays made by templated solid-state dewetting. *Small*, 5:860, 2009.
14. L. Rayleigh. On the instability of jets. *Proc. Lond. Math. Soc.*, page 4, 1878.
15. Donghyun Kim, Amanda L. Giermann, and Carl V. Thompson. Solid-state dewetting of patterned thin films. *Appl. Phys. Lett.*, 95:251903, 2009.

Chapter 5
Self-Organized Nanoparticle Arrays:
Optical Properties

In this chapter, we present the optical characterizations of the specimens during all the steps of the fabrications, from the bare LiF substrates to the 2D arrays of gold nanoparticles.

5.1 Optical Properties of Lithium Fluoride Substrates

In this section we will focus on the characterization of the optical properties of the LiF substrates. As will become more clear in the next chapter, they will play an important role for the modelling of the optical response of the gold NP.

Lithium fluoride is a well-known material for dosimetry and optoelectronics. It has a very wide band gap of 13.6 eV, so in the visible and near-IR range it behaves like a transparent, non absorbing material (transmittance higher than 90 % in the range 0.1 –7 μm [1, 2]). Its optical characteristics have been extensively investigated mainly in connection with its color centers [3–6]. These, usually induced by oxygen and metal ions impurities, introduce weak absorptions in the 200–350 nm region [7, 8], conferring the crystal a characteristic yellowish color.

We first consider the LiF(110) samples in the as-received state, whose morphology has been reported in Fig. 4.2. Macroscopically, they appear completely transparent and uncoloured, without any evident absorption. We report in Fig. 5.1 the typical transmission (left panel) and ellipsometric (right panel) spectra for such samples, measured in the 280–1,700 nm range. Starting from lower energies, the transmittance (panel a) has an almost constant value of 0.95 up to $E \approx 3$ eV (corresponding to the wavelength range $\lambda = 400$–1, 700 nm), and then slowly decreases approaching the UV region. No significant spectral features are visible in the 3.5–6 eV (200–350 nm) range, characteristic of the colour centers, confirming the good quality of the crystals. Panel b shows Ψ and Δ spectra measured at an angle of incidence $\theta = 50°$. The values of Δ are very close to 180°, the expected value for an ideally non-absorbing dielectric [9]. Starting from $\lambda \approx 800$ nm and proceeding to lower wavelengths,

L. Anghinolfi, *Self-Organized Arrays of Gold Nanoparticles*, Springer Theses,
DOI: 10.1007/978-3-642-30496-5_5, © Springer-Verlag Berlin Heidelberg 2012

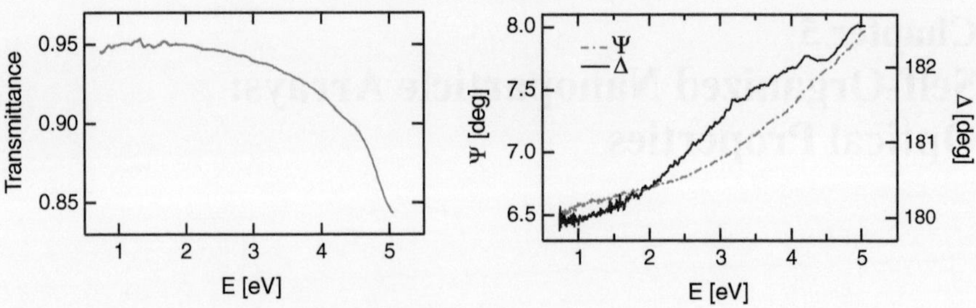

Fig. 5.1 Panel (a): transmission spectrum measured at normal incidence on a double-side polished LiF(110) sample in the as received state. Panel (b): Ψ (*red curve*) and Δ (*black curve*) spectra measured on a single-side polished LiF(110), at incidence $\theta = 50°$

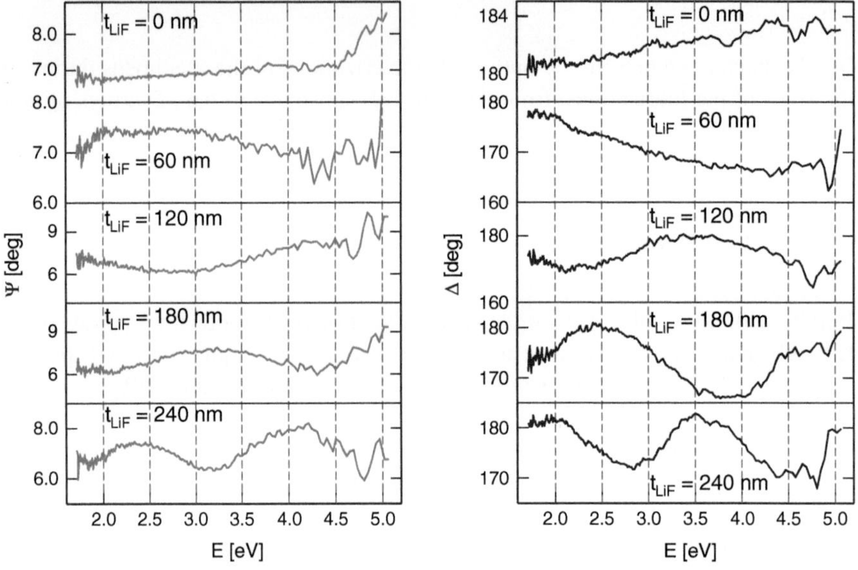

Fig. 5.2 Ψ (*left panel*) and Δ (*right panel*) spectra at incidence angle $\theta = 50°$, measured during the fabrication of a nanopatterned LiF(110) sample (growth temperature $T = 300\,°C$, periodicity $\Lambda = 35\,nm$), as a function of the deposited thickness t_{Lif}

Δ tends to become slightly larger than 180°, probably indicating the presence of a thin surface layer having an optical response slightly different from the bulk crystal.

Since the LiF(110) is a morphologically anisotropic surface (Fig. 4.1d), we investigated the optical behaviour of the LiF substrates varying the orientation of the plane of incidence, with respect to the crystalline axes. Within the experimental accuracy, we found no significant variations of the ellipsometric response, either setting the plane of incidence along the [001] or the [1$\bar{1}$0] directions, indicating no influence of the crystal symmetry on the optical properties of LiF.

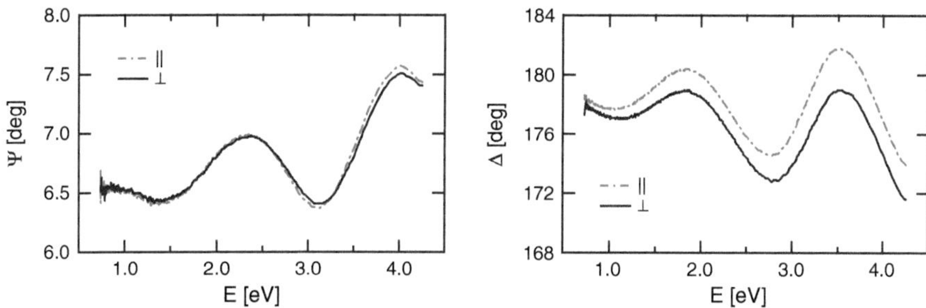

Fig. 5.3 Ψ (*left panel*) and Δ (*right panel*) spectra, measured at an angle of incidence $\theta = 50°$, of a nanopatterned LiF(110) sample, with periodicity $\Lambda = 35\,\text{nm}$, for different orientations of the sample with respect to the optical plane. *Red line*: plane of incidence parallel (‖) to the ripples direction. *Black line*: plane of incidence transverse ($\perp$) to the LiF ridges

In analogy with the morphological measurements, we checked the evolution of the LiF optical response, monitoring the ellipsometric spectra during the LiF homoepitaxy as a function of the amount of deposited LiF. In Fig. 5.2 we show Ψ and Δ spectra for a sample with periodicity $\Lambda \approx 35\,\text{nm}$ (Fig. 4.4); like in Fig. 4.4, the curves were measured ex situ at an angle of incidence $\theta = 50°$, at steps of 60 nm of deposited LiF during the deposition of a 240 nm thick film. Increasing the thickness of the deposited LiF, we notice that the average values of the Ψ and Δ spectra remain almost unchanged, while the main variations consist in the appearance of weak oscillations in the spectra, which become more closely spaced at increasing LiF thickness. These oscillations are due to the interference between the multiple reflections inside the deposited film; in particular, their periodicity is directly related to the thickness of the film, making spectroscopic ellipsometry an effective technique to monitor ex situ the LiF homoepitaxy and to calibrate with a precision of few nanometers the deposition rate of the effusion cell.

Another characteristic feature induced by the deposition of LiF is a weak but apparent optical anisotropy between measurements performed with the optical plane parallel to the [001] and [1$\bar{1}$0] directions. This is shown in Fig. 5.3, where we report Ψ and Δ spectra measured at $\theta = 50°$ with the plane of incidence oriented either parallel (red lines) or perpendicular (black lines) to the ripple direction; the anisotropy is more pronounced for Δ, where it is as great as $3°$ at the highest energies, while it is barely noticeable for Ψ. Since it is not observed for the bare substrate, we conclude that it must be induced by the deposited film, and in particular by the intrinsic morphological uniaxial asymmetry of the ripple structure. As is often the case in ellipsometry, the direct interpretation of the spectra, and of the corresponding anisotropy, is not a straightforward task, and an analytical modelling is typically required. We will discuss these effects in the next chapter (Sect. 6.1).

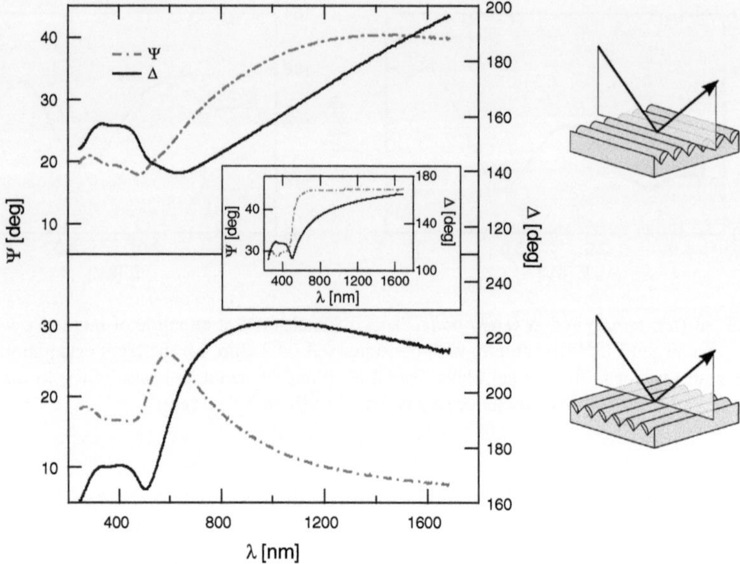

Fig. 5.4 Ellipsometric spectra acquired at an angle of incidence $\theta = 50\,°$ measured on a nanopatterned LiF(110) sample after the deposition of $t_{Au} = 5\,$nm Au at $T \approx 100\,°$C, with the plane of incidence oriented either along (*top panel*) or across (*bottom panel*) the ripples. Inset of the figure: Ψ and Δ curves ($\theta = 50°$) for a bulk gold sample

5.2 Optical Properties of Gold Nanoparticles Arrays

We will now address the optical response of the arrays of gold nanostructures. As seen in Sect. 2.3, metallic structures with a typical size of few tens of nanometers can sustain characteristic plasmonic resonances, whose properties depend on the specific shape and environment of the nanoparticles. Here, by tuning the fabrication of the arrays, we explore the variations in the plasmonic response as a function of the systems' characteristics.

5.2.1 Gold Nanowires

We start by considering the optical properties of gold nanostructures realized directly following the grazing-angle deposition onto the rippled LiF substrates without any post-deposition treatment. In Fig. 5.4 we report Ψ and Δ spectra measured on the same sample described in Fig. 4.7d. In analogy with the previous case, the plane of incidence was set either parallel or perpendicular to the ripples direction. In the inset of the figure, the corresponding Ψ and Δ spectra measured for a reference bulk gold sample are also reported.

Comparing Fig. 5.4 with Fig. 5.3, we notice that the ellipsometric spectra have drastically changed after the deposition of Au, and that the contribution of gold to the optical spectra has clearly become dominant. When the plane of incidence is along the ripples (top panel of Fig. 5.4) Ψ has a very broad maximum at $\approx 1,300$ nm, while Δ monotonically increases, almost linearly with increasing wavelength; on the contrary, when the plane of incidence is rotated by $90°$, Ψ is sharply peaked around 600 nm and Δ shows a broad maximum at $\approx 1,000$ nm. The main feature is now a strong in-plane optical birefringence, that becomes most pronounced in the wavelength range between 600 and 1,700 nm. Furthermore, despite some similarities at the lowest wavelengths, these curves are also markedly different compared to the bulk reference spectra (inset of Fig. 5.4). From purely qualitative considerations, we can assign the main features of the optical response of the system to the occurrence of electron confinement within the gold nanostructures.

Although spectroscopic ellipsometry is probably the most powerful optical method for the characterization of materials, in this particular case it might not represent the most straightforward manner to achieve a clear and simple physical picture of the phenomena occurring within the Au nanostructures. The onset of plasmonic response in a system is in fact typically manifested with the appearance of a sharp reflectivity or absorption feature in the spectra. When measuring a ratio of reflectivities, as done in ellipsometry, such sharp features typically tend to compensate each other, thereby effectively cancelling out the most remarkable manifestation of plasmonic response. For example, the peak observed in Ψ for the plane of incidence perpendicular to the ripples (lower panel in Fig. 5.4) does not correspond to a "true" plasmonic resonance, but simply arises as the result of the ratio between the p and s reflectivities [see (2.25)]. In addition to this, we also notice that the resonances are more easily interpreted as enhancements of light reflectivity or absorption, rather than phase variations. A more direct approach to investigate the plasmonic properties is therefore to measure separately the plain p and s reflectivities or transmissivity, and analyse the spectra features that appear there.

For simplicity we keep the same configurations used in the ellipsometric measurements, and we define a parallel ($\parallel$) and a perpendicular ($\perp$) geometry for the plane of incidence respectively parallel and perpendicular the ripples direction, as sketched in Fig. 5.5 for reflection. In parallel geometry (panel (a)), when the incident light is s-polarized the electric field is purely transverse to the ripples, oscillating in the $[1\bar{1}0]$ direction, while for p-polarized light, it has a component both along the ripples and normal to the surface. In perpendicular ($\perp$) geometry (Fig. 5.5b), instead, the system is excited purely parallel (s-polarized light) or transverse (p-polarized light) to the ripples, in the latter case simultaneously along the in-plane $[1\bar{1}0]$ and out-of-plane $[110]$ directions. We will use the following notation when referring to such reflectivities: $R_S^{\parallel}$, $R_P^{\parallel}$, $R_S^{\perp}$, $R_P^{\perp}$.

In Fig. 5.6 we report sets of reflectivities recorded at an angle of incidence of $50°$, measured on various samples after the deposition of $t_{Au} \approx 5$ nm of gold. The blue curves were measured on the same sample described in Fig. 4.7b, having periodicity $\Lambda \approx 45$ nm, while the black and red ones correspond to the samples in Fig. 4.7c

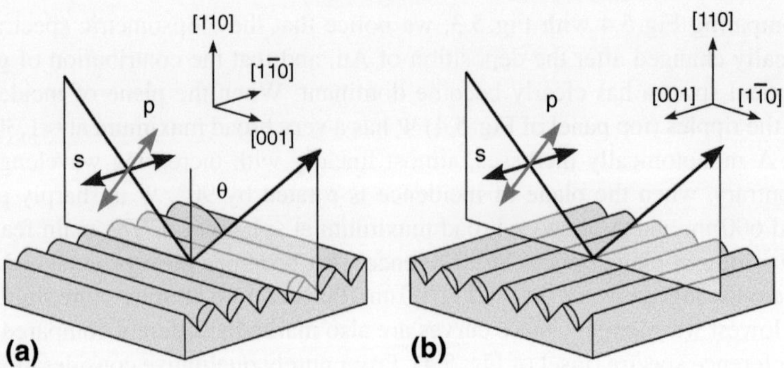

Fig. 5.5 Optical geometries for the reflectivity measurements, and corresponding orientations of the electric fields for either *s*- or *p*-polarized incident light. Plane of incidence parallel (**a**) and perpendicular (**b**) to the ripples ridges

(with $\Lambda \approx 25$ nm) and in Fig. 4.7d (with $\Lambda \approx 40$ nm), respectively. The reflectivity of bulk gold is also reported in Fig. 5.6c as a reference. Furthermore, in Fig. 5.7a we show the transmission spectra for the sample with $\Lambda \approx 25$ nm, corresponding to the black curves of the previous graphs. The spectra were measured at normal incidence, with light polarized either parallel (red curve) and perpendicular (black curve) to the ripples.

In full analogy with ellipsometric data, a strong optical anisotropy is observed also in the reflectivity/transmission spectra, when the electric field oscillates either perpendicular or parallel the ripple direction: in the first case ($R_S^{\parallel}$ and $R_P^{\perp}$) the reflectivities are peaked at $\lambda \approx 600$ nm and then decrease at longer wavelengths; at variance with this, $R_S^{\perp}$ and $R_P^{\parallel}$ have a higher intensity, with a very broad maximum around 1,000 nm. Both of these trends are quite different from the bulk reference, where the reflectivity approaches unity for $\lambda \gtrsim 600$ nm. Considering the transmission, the same features observed as enhanced reflectivity are observed as minima of the transmitted intensity: for light polarized along the ripples, the transmittance is as low as 0.4 over a large region from 800 nm to the largest wavelength (cfr. the transmittance of the bare substrate in Fig. 5.1a); when the electric field is instead perpendicular to the ripples, the overall transmitted intensity is higher, but shows a sharp minimum at $\lambda \approx 650$ nm.

With the support of the AFM data, we can safely assign the peaks of reflectivity and the minima of transmittance to the occurrence of plasmonic resonances inside the gold grains. In the direction perpendicular the ripples, the grains are well separated by the ripples ridges, so the electrons are effectively confined and the LSP peaks quite sharp. On the contrary, inside each groove the Au grains can accommodate as close as possible to one another, and there are no constraints on their dimensions; the broadening of the resonances, when the electric field is parallel the ripples, can then be attributed in part to the dispersion of the grains lengths, so that the observed peak is actually the superimposition of several peaks slightly shifted in wavelength,

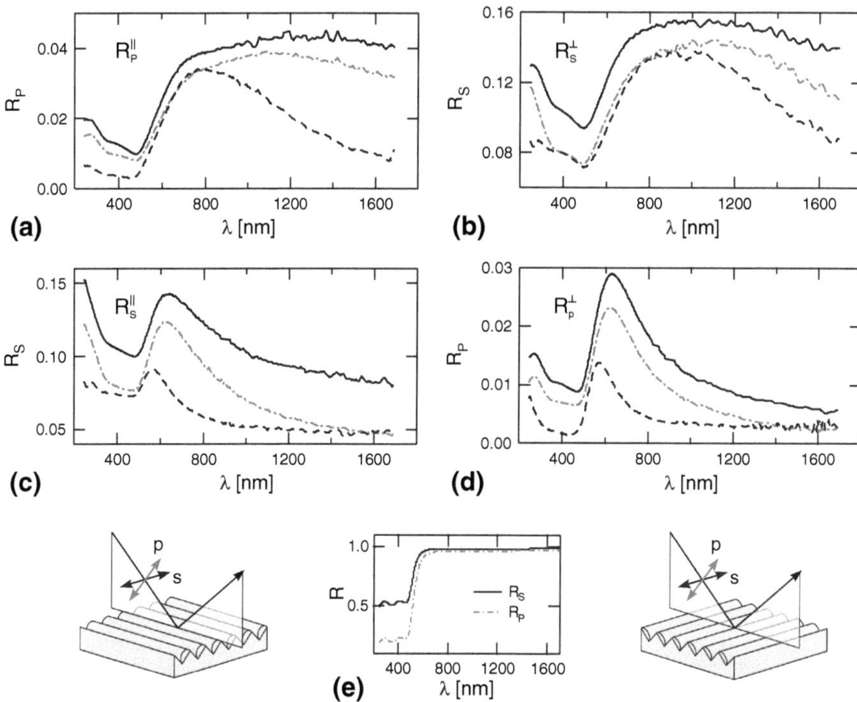

Fig. 5.6 Sets of reflectivities at an angle of incidence $\theta = 50°$ for s- and p-polarized incident light, measured on various gold "nanowires" samples ($t_{Au} \approx 5$ nm) on nanopatterned LiF(110). *Black continuous curves*: $\Lambda = 45$ nm; *red dot-dashed curves*: $\Lambda = 40$ nm; *blue dashed curves*: $\Lambda = 25$ nm. Plane of incidence either parallel (*left panels*) or perpendicular (*right panels*) to the LiF ridges. Exciting electric field along (panels a, b) and transverse (panels c, d) to the gold wires. Panel e: s- and p-polarized reflectivities of bulk gold

in part to the coalescence of several contiguous grains, which reduces the electronic confinement restoring a metallic behaviour somewhat analogous to a bulk sample. Thus, we will refer to this class of samples as gold "nanowires".

Having measured the reflectivities, we can now give a qualitative interpretation of the Ψ spectra, which are roughly equivalent to the ratio R_P/R_S between the reflectivities; we consider only the region of interests $\lambda \gtrsim 500$ nm. When the plane of incidence is perpendicular to the ripples, Ψ (Fig. 5.4b) corresponds to the ratio between $R_P^{\perp}$, peaked at $\lambda \approx 600$ nm, and $R_S^{\perp}$, slowly varying, so it has a sharp peak similar to $R_P^{\perp}$ but shifted at a different wavelength, because $R_S^{\perp}$ is not a constant. On the contrary, Ψ as measured in $\parallel$ geometry (Fig. 5.4a) is roughly the ratio between a nearly constant curve ($R_P^{\parallel}$) and a curve peaked at $\lambda \approx 600$ nm ($R_S^{\parallel}$), resulting as broad as $R_P^{\parallel}$, but with a lower shoulder around 600 nm. This qualitative analysis shows that the plasmonic resonances as seen by ellipsometry are difficult to interpret, because they appear either shifted in wavelength, as in $\perp$ geometry, or completely

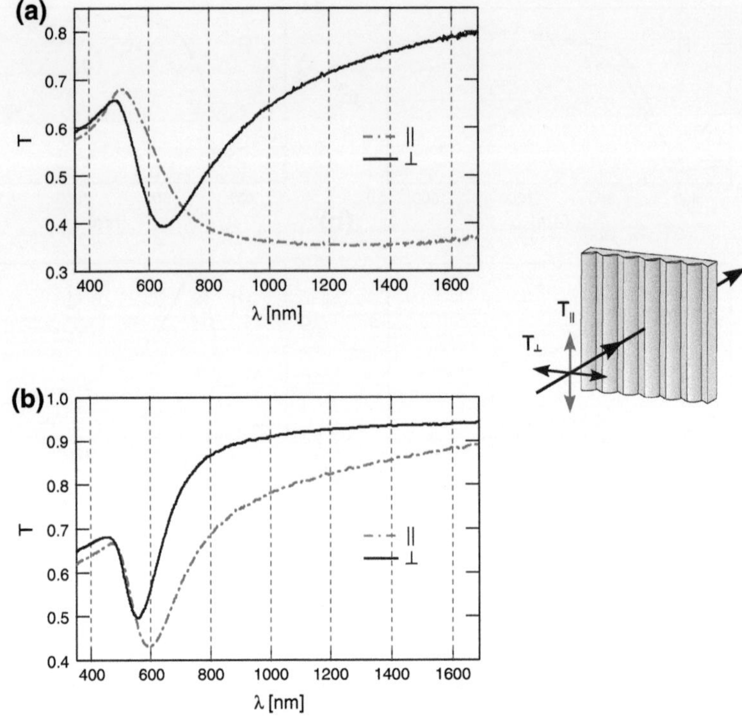

Fig. 5.7 Transmissivity spectra measured at normal incidence on a nanopatterned LiF(110) ($\Lambda \approx$ 25 nm) after the deposition of $t_{Au} \approx 5$ nm gold at $T = 100\,°C$ (panel **a**) and subsequent annealing at $T = 400\,°C$ (panel **b**). The transmissivities were acquired with linearly polarized light, parallel (*red curves*) and perpendicular (*black curves*) to the LiF ridges

hidden, as in ∥ geometry. Therefore, in order to investigate the plasmonic response of the gold nanostructures and have a clear picture of the resonances, from now on we will concentrate on reflectivity measurements rather than ellipsometry.

5.2.2 From Au Nanowires to Au Nanoparticles

In Sect. 4.2 we showed that the formation of 2D arrays of gold nanoparticles is induced by a mild thermal annealing of the gold nanowires. Using the setup shown in Fig. 3.3, we could monitor in situ the evolution the plasmonic peaks during the annealing procedure. To this end, in Fig. 5.8a we report a set of $R_S^{\parallel}$ spectra acquired for a sample with $\Lambda \approx 55$ nm and $t_{Au} \approx 7$ nm (Fig. 4.8d), as a function of the annealing temperature in the range between 50 and 400 °C (the angle of incidence was fixed at 60°). Increasing the temperature, the resonance progressively blue shifts and gets narrower. The position of the LSP peak as a function of T is shown in

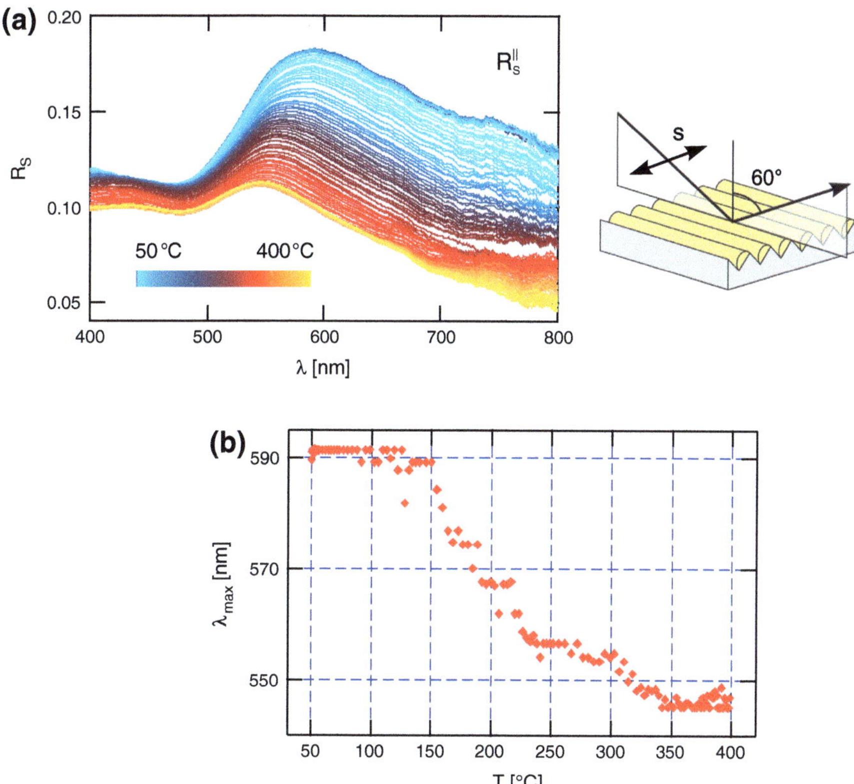

Fig. 5.8 Panel **a** set of $R_S^{\|}$ spectra (s-polarized incident light and plane of incidence along the ripples) at incidence $\theta = 60°$, measured in situ during the annealing at $T = 400\,°C$ of a gold nanowires sample. Panel **b** evolution of the LSP peak wavelength during the annealing, as a function of the temperature

panel (b). The position of the resonance remains almost unchanged up to $\approx 150\,°C$, then it rapidly shifts to lower wavelengths until $T \approx 250\,°C$ is reached, and lastly blue-shifts slightly again between 300 and 350 °C.

A direct comparison between the reflectivities before and after the annealing is shown in Fig. 5.9 for the sample of Fig. 4.8c, characterized by $\Lambda \approx 25\,nm$ and $t_{Au} \approx 5\,nm$; the curves were acquired ex situ, the red ones prior to the annealing, the black ones afterwards. In panels (c) and (d) we report the reflectivities $R_S^{\|}$ and $R_P^{\perp}$, similar to the geometry of Fig. 5.8 except at an angle of incidence of 50°. According to the observations made for Fig. 5.8 following the annealing, the LSP peak, in the direction perpendicular the ripples, slightly blue shifts and the resonance become sharper. The main effects of annealing, however, are visible in panels (a) and (b) of Fig. 5.9, in $R_P^{\|}$ and $R_S^{\perp}$ geometries, where the electric field is mostly oriented along the ripples. Here, the very broad peak observed for the nanowires is replaced,

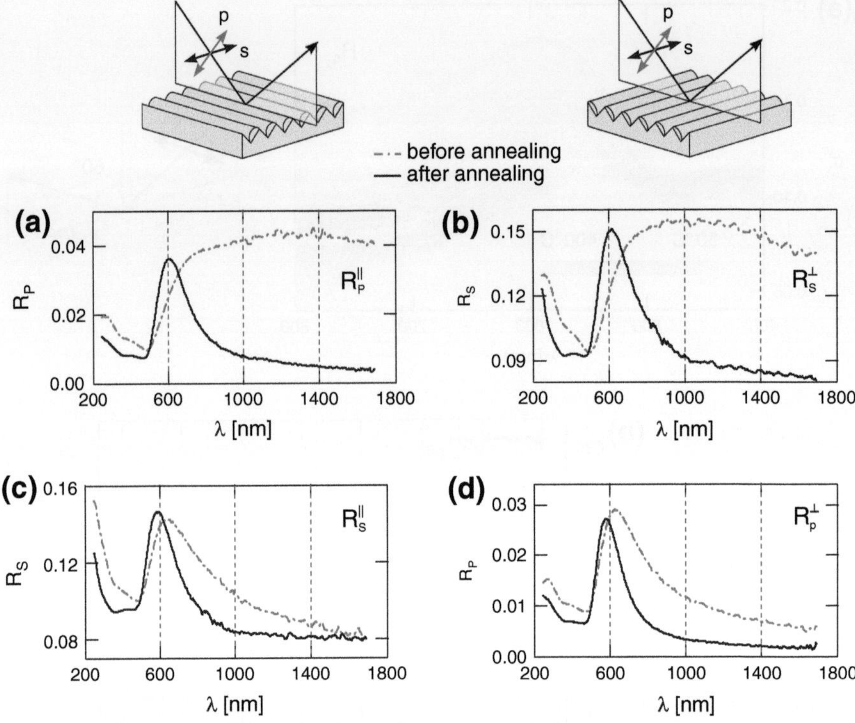

Fig. 5.9 Reflectivities, at an angle of incidence $\theta \approx 50°$, measured before (*red curves*) and after (*black curves*) annealing at $T = 400\,°C$ of gold nanowires on LiF(110) ($\Lambda = 25\,nm$). Plane of incidence parallel (*left panels*) and perpendicular (*right panels*) to the ripples. EM excitation along (panels **a**, **b**) and across (panels **c**, **d**) the ripples

after the annealing, by a much narrower peak, shifted of $\lambda \approx 400\,nm$ towards lower wavelengths.

The corresponding comparison, concerning the same sample, between the transmissivities before and after the annealing is instead reported in Fig. 5.7; the spectra were measured at normal incidence, with the electric field oriented either parallel ($T_\parallel$, red curve) or perpendicular ($T_\perp$, black curve) to the LiF ridges. Similarly to what observed in reflection, the broad minimum observed in $T_\parallel$ after the Au deposition has now evolved into a sharp absorption at $\lambda = 600\,nm$, while the minimum in $T_\perp$ blue shifts from $\lambda \approx 650\,nm$ to $\lambda \approx 560\,nm$ and gets narrower. We notice that the positions of the resonances when observed in transmission or in reflection are slightly different, the peaks of R being systematically red shifted with respect to the minima of T; we will discuss these effects in Sect. 6.2.2.

We can interpret the variations in the optical properties induced by the annealing procedure using the same arguments of the previous sections. We saw in Sect. 4.2 that, following the annealing, the layer of gold reorganizes in ordered arrays of nanoparticles; however, due to the convolution of the AFM tip, we could not determine from

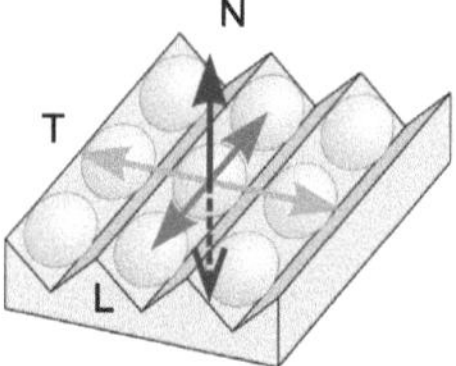

Fig. 5.10 Schematic representation of the three different LSP modes for the gold nanoparticles. Longitudinal mode (*L*): EM excitation along the ripples. Transverse mode (*T*): excitation transverse to the LiF ridges. Normal mode (*N*): excitation normal the plane of the sample

the AFM data whenever the particles were disconnected or not. In this respect, reflectivity and transmissivity provide complementary tools to characterize the NP arrays. In the previous section we associated the presence of a sharp LSP resonance, in the direction perpendicular the ripples, to the formation of disconnected wires of gold. In the same way, we can explain the sharp resonance, observed when the electric field is applied along the ripples, as the result of the formation of isolated metallic nanostructures within each groove: comparing these considerations with the AFM data, we can then confirm that the nanoparticles of gold are disconnected both along and perpendicular to the ripples.

5.2.3 Gold Nanoparticles

We now focus on the analysis of the optical properties of the 2D arrays of nanoparticles. In particular, we separately consider the dependence of the plasmonic resonances on the ripples periodicity and on the gold coverage. For simplicity, we identify 3 distinct plasmon modes, related to the geometry of the system (Fig. 5.10): a *longitudinal* (L) mode is excited when the electric field oscillates along the ripples; a *transverse* (T) mode is excited when the electric field is applied across the ripples in the plane of the sample; a *normal* (N) mode is excited when the electric field has a component normal the sample surface. We will discuss in the next chapter the validity of this assumption. Considering the reflectivity geometries introduced in the previous sections, we can deduce that $R_S^{\parallel}$ and $R_S^{\perp}$ geometries allow to excite the T and L modes only, respectively, while in $R_P^{\parallel}/R_P^{\perp}$ geometries the N mode is activated beside the L/T mode. However, as we will see in the next sections, the intensity of the N mode is typically much weaker than the L and T ones, therefore the corresponding contribution to the reflectivity or absorption peaks remains always "hidden" by the other optical features, and only the L and T modes are clearly observed in reflection or transmission.

Following the same notation of the plasmonic modes, we define the length, the width and the height of the particles as the dimensions along and across the ripples and normal to the surface, respectively.

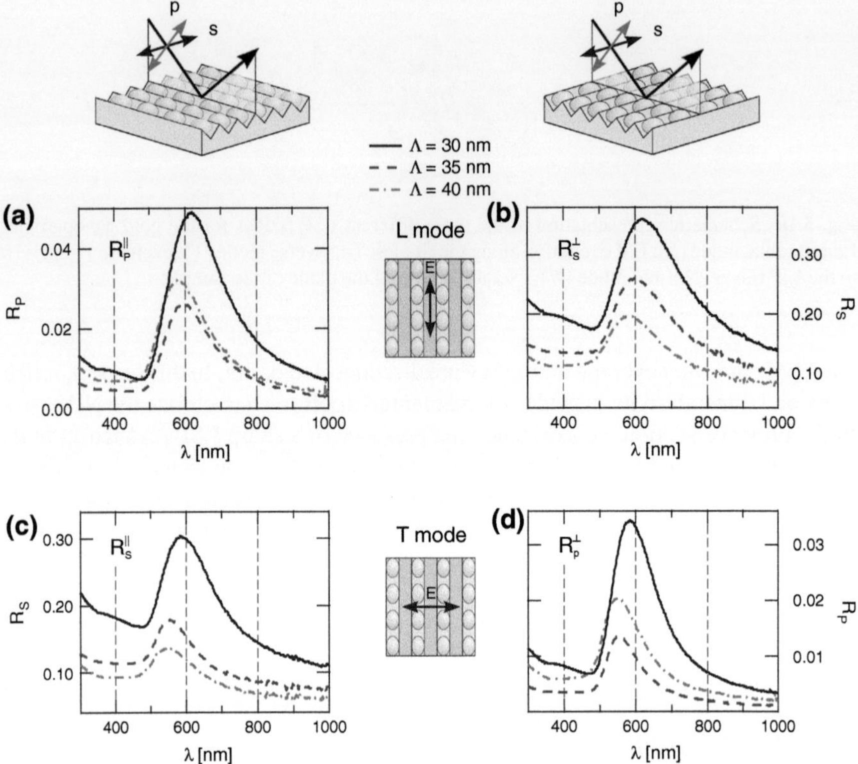

Fig. 5.11 Sets of reflectivities at $\theta = 50°$ incidence, for s- and p-polarized incident light, measured on nanopatterned LiF(110) samples with different periodicities Λ, after the deposition of $t_{Au} = 5$ nm Au at $T \approx 100\,°C$ and annealing at $T = 400\,°C$. Plane of incidence parallel (*left panels*) and perpendicular (*right panels*) the ripples. Excitation of LSP L mode (panels **a, b**) and T mode (panels **c, d**)

Dependence of LSP Modes on Ripple Periodicity

We first investigate the influence of the ripples periodicity on the position of the plasmonic resonances. In Fig. 5.11 we report sets of reflectivities measured on several arrays of in-plane spherical particles at 50° of incidence. In all cases $\approx$5 nm of gold were deposited, and the samples have different periodicities of the underlying ripple structure: $\Lambda \approx 30$ nm for black lines (sample in Fig. 4.8a), $\Lambda \approx 35$ nm for blue lines, and $\Lambda \approx 40$ nm for red lines (sample in Fig. 4.8b).

For decreasing Λ, we observe two main variations of the resonances: both the L and T modes shift towards longer wavelengths, the L mode shifting from $\lambda \approx 580$ nm for $\Lambda = 30$ nm to $\lambda \approx 620$ nm for $\Lambda = 40$ nm, and the T mode from $\lambda \approx 550$ nm to $\lambda \approx 585$ nm; correspondingly, the intensity of the reflectivities increases, and for $\Lambda = 30$ nm is nearly twice as much as for Λ=40 nm.

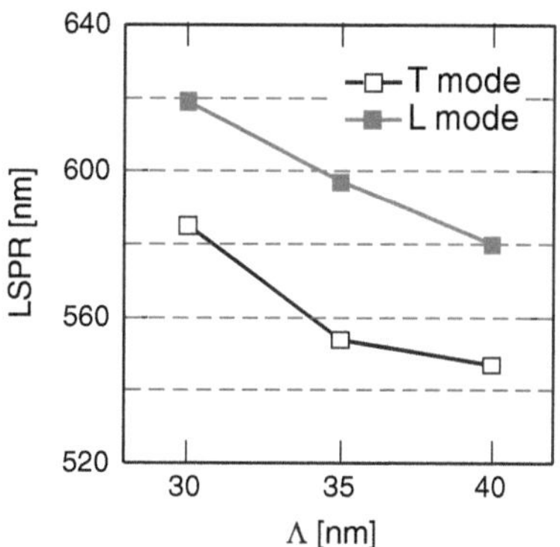

Fig. 5.12 Wavelength position of longitudinal (*solid red squares*) and transverse (*open black squares*) LSP resonances for gold nanoparticles on LiF(110) samples with different ripples periodicities Λ and the same amount $t_{Au} = 5\,$nm of deposited gold (see Fig. 5.11)

We can qualitatively explain these behaviours by considering the NP particles spacings and interactions strength. As seen in the previous chapter Sect. 4.2, the ripples structure fixes both the NPs separation in the direction across the ripples, equal to Λ, and the width of the NPs; instead, the center-to-center distance between adjacent NPs belonging to the same groove depends on the lengths of the NPs. However, since the particles in Fig. 5.11 have an in-plane circular section, in these cases the ripples periodicity Λ also affects the longitudinal separation, so when Λ is reduced the particles grow smaller and get closer in all directions. This has two main consequences: by decreasing the NPs spacings, the mutual interactions become stronger, leading to the observed red shift of both the L and T plasmon modes; moreover, the enhancement of the local electric fields promotes stronger induced dipoles, only partially compensated by the reduction of NP volumes, and accordingly a higher reflectivity.

The position of the LSP resonances as a function of the periodicity Λ is summarized in the graph in Fig. 5.12. The asymmetry of the L and T modes, despite the in-plane spherical shape of the particles, is due to the different NPs spacings in the directions along and across the ripples. In fact, the particles are more packed inside the individual grooves rather than across the ripples, so the EM interactions between the particles are stronger when the electric field is parallel the ripples ridges: this leads to a relative red shift of the L modes with respect to the T modes.

Dependence of LSP modes on Au thickness

In this section we briefly review the effects of the thickness t_{Au} of the deposited gold on the LSPs, keeping the ripple periodicity Λ fixed. We report in Fig. 5.13 sets of reflectivities measured on different arrays of NPs. The arrays were grown on substrates with a similar periodicity of $\Lambda \approx 45\,$nm, increasing the Au thickness from $t_{Au} \approx 4.5\,$nm (red curves, sample in Fig. 4.8b), to $t_{Au} \approx 5.4\,$nm (blue curves, sample in Fig. 4.8c), to $t_{Au} \approx 7.2\,$nm (black curves, sample in Fig. 4.8d). The corresponding positions of the LSP peaks as a function of the Au thickness is reported in Fig. 5.14.

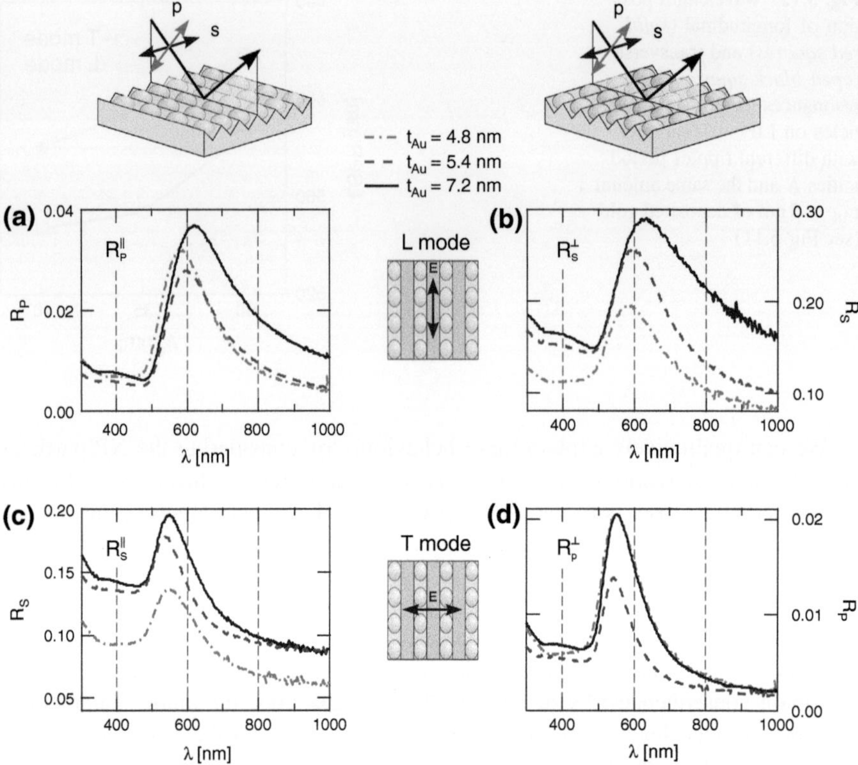

Fig. 5.13 Sets of reflectivities at $\theta = 50°$ incidence, for s- and p-polarized incident light, measured on nanopatterned LiF(110) samples with similar periodicities $\Lambda \approx 45$ nm, after the deposition of different amounts of gold at $T \approx 100\,°C$ and annealing at $T = 400°$. Plane of incidence parallel (*left panels*) and perpendicular (*right panels*) to the ripples. Excitation of LSP L mode (panels **a**, **b**) and T mode (panels **c**, **d**)

Comparing the spectra, we observe variations of the optical response somewhat similar to the previous case. The intensity of the reflectivities increases with increasing t_{Au}, and correspondingly the L mode red-shifts; the T mode, instead, remains almost unchanged.

In analogy with the previous case, we can interpret qualitatively these effects. As described in the previous chapter, the thickness of the deposited gold affects the shape of the individual NPs; in this case we are dealing with arrays of particles characterized by different aspect ratios, gradually varying from in-plane spherical to elongated (see Fig. 4.8). However, since the arrays were grown on equivalent substrates, the NPs have similar widths and separations in the direction across the ripples, the main differences being the lengths and the spacings along the ripples. Recalling the arguments in Sect. 2.3.3, we can therefore associate the splitting of the LSP resonances mainly to geometrical effects: increasing t_{Au}, the NPs aspect ratio

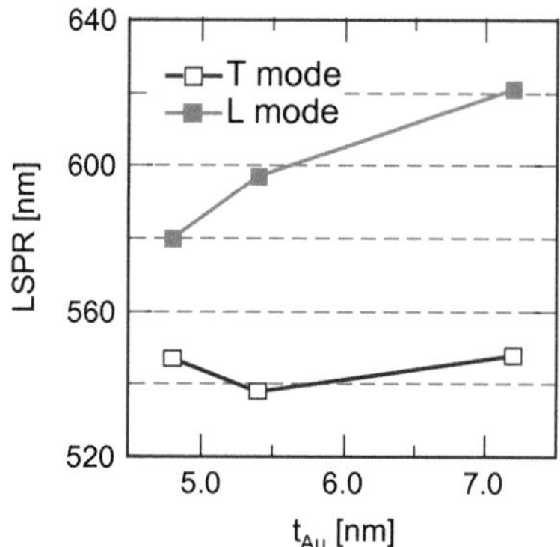

Fig. 5.14 Wavelength position of longitudinal (*solid red squares*) and transverse (*open black squares*) LSP resonances for gold nanoparticles on LiF(110) samples with similar ripples periodicities and the different amounts t_{Au} of deposited gold (see Fig. 5.13)

increases, inducing a red-shift of the L mode and a slight blue shift of the T mode (Fig. 2.15). Instead, the increase of reflectivity can be related mainly to the overall increase of the NPs volume, which promotes more intense induced dipoles ($\alpha \propto v$, see (2.42)) and consequently an enhanced optical response.

References

1. H. H. Li. Refractive index of alkali halides and its wavelength and temperature derivatives. *J. Phys. Chem. Ref. Data*, 5:329, 1976.
2. M. Bass, C. DeCusatis, G. Li, V.N. Mahajan, E.V. Stryland, and C. MacDonald. *Handbook of Optics: Optical properties of materials, nonlinear optics, quantum optics*. Handbook of Optics. McGraw-Hill, 2009.
3. Rosa Maria Montereali, Antonella Mancini, Giancarlo C. Righini, and Stefano Pelli. Active stripe waveguides produced by electron beam lithography in LiF single crystals. *Opt. Comm.*, 153:223, 1998.
4. Kenichi Kawamura, Masahiro Hirano, Toshio Kurobori, Daizyu Takamizu, Toshio Kamiya, and Hideo Hosono. Femtosecond-laser-encoded distributed-feedback color center laser in lithium fluoride single crystals. *App. Phys. Lett.*, 84:311, 2004.
5. G. Baldacchini, M. Cremona, G. d'Auria, R. M. Montereali, and V. Kalinov. Radiative and nonradiative processes in the optical cycle of the F_3^+ center in LiF. *Phys. Rev. B*, 54:17508, 1996.
6. R. M. Montereali, M. Piccinini, and E. Burattini. Amplified spontaneous emission in active channel waveguides produced by electron-beam lithography in LiF crystals. *App. Phys. Lett.*, 78:4082, 2001.
7. G. Baldacchini, O. Goncharova, V. S. Kalinov, R. M. Montereali, A. Vincenti, and A. P. Voitovich. Thermal transformation of colour centres in lif crystals with given content of oxygen, hydroxyl and metal ions. *Phys. Stat. Sol. (c)*, 4:1134, 2007.
8. G. Baldacchini, O. Goncharova, V. S. Kalinov, R. M. Montereali, E. Nichelatti, A. Vincenti, and A. P. Voitovich. Optical properties of coloured lif crystals with given content of oxygen, hydroxyl and metal impurities. *Phys. Stat. Sol. (c)*, 4:744, 2007.
9. H. Fujiwara. *Spectroscopic Ellipsometry. Principles and Applications*. Wiley, 2007.

Chapter 6
Modelling and Analysis of the Optical Properties

In this chapter we will focus on the development of a theoretical modelling framework, allowing us to account for the optical response of the 2D arrays of gold nanoparticles described in the previous chapters. A theoretical support to the experimental data is of fundamental importance in order to achieve a comprehensive understanding of the origins of the optical properties of these systems, and thus to engineer the optical response by selecting a priori the proper morphological characteristics.

In the next sections, we will first characterize the optical properties of the LiF templates, and then we will cover in detail the arrays of gold NPs, overlooking instead on the nanowires.

6.1 Lithium Fluoride Nanostructures

We start by extracting the optical constant of the nanopatterned LiF substrates. A detailed characterization of the substrates characteristics is in fact extremely important for a proper modelling of the Au NPs arrays. In Sect. 5.1 we reported the ellipsometric spectra measured on the LiF(110) substrates prior to and after the further deposition of LiF, so here we proceed, as customary in ellipsometry, by constructing a model to describe the samples and then fit the optical constants against the experimental data. All the calculation have been performed using the WVASE software supplied with the ellipsometer.

6.1.1 LiF Substrates

We first consider the LiF(110) substrates in the as-received state. In the previous chapter we saw that the samples exhibit negligible in-plane birefringence, despite the anisotropic atomic environments along the [001] and [1$\bar{1}$0] crystal directions.

L. Anghinolfi, *Self-Organized Arrays of Gold Nanoparticles*, Springer Theses, DOI: 10.1007/978-3-642-30496-5_6, © Springer-Verlag Berlin Heidelberg 2012

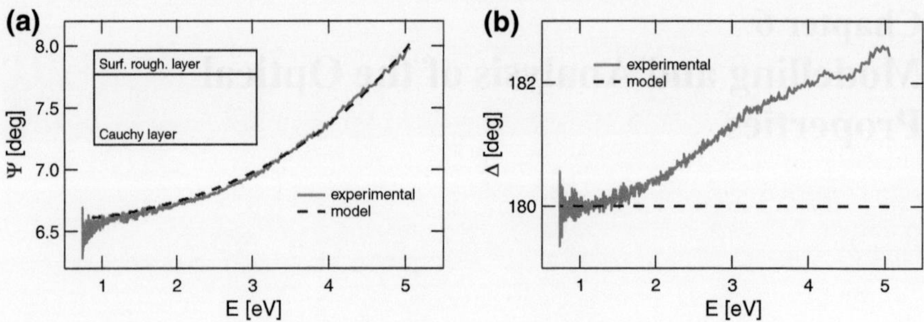

Fig. 6.1 *Red lines* Ψ (*left panel*) and Δ (*right panel*) spectra measured at incidence $\theta = 50°$ on LiF(110) crystals in the as-received state. *Black lines* corresponding calculated curves, fitted using the model sketched in the inset

Fig. 6.2 Real part of the refractive index for a Cauchy layer representative of bulk LiF, as obtained from the fits in Fig. 6.1; the corresponding imaginary part is null at all wavelengths

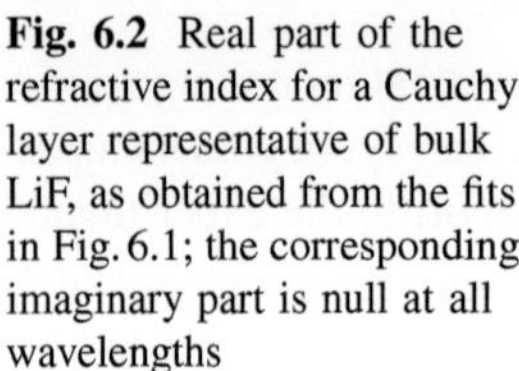

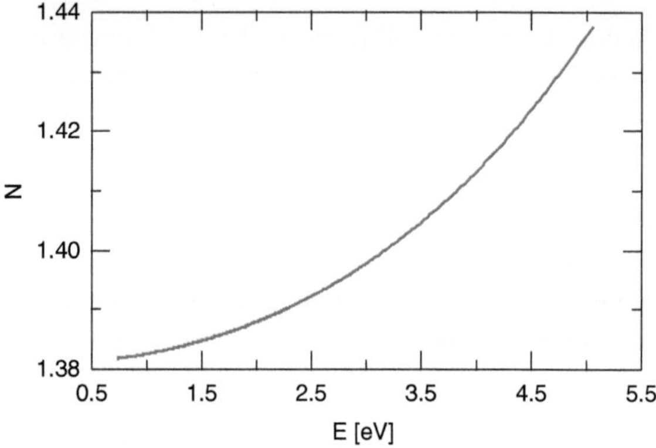

We can therefore apply an isotropic dielectric model for the bare substrate. Lithium fluoride is an insulator with a very wide band gap, and in the energy range between ~ 1 and ~ 5 eV it behaves like a transparent, non absorbing material; then, we can approximate its optical constants, in the visible and near-IR range, using a Cauchy parametrization of the refractive index [see Eq. (2.20)]. We can further improve the model by accounting for the surface roughness; in fact, AFM analysis of the substrates revealed a mean RMS roughness of about 6–9 nm (see Sect. 4.1, Fig. 4.2), and, given the high sensibility of ellipsometry, this is also expected to affect the spectra, mostly Δ. Since roughness is a standard element in ellipsometric analysis, the software already provides the appropriate methods to account for it.

A sketch of the overall model and the corresponding fit results for Ψ and Δ are reported in Fig. 6.1, where the optical constants of the substrate and the thickness of the roughness layer were employed as free fit parameters. The red curves in figure correspond to the spectra measured at $\theta = 50°$ of incidence, while the fits are plotted as black lines. The real part N of the best-fit refractive index is shown in Fig. 6.2, the imaginary part is instead null at all the wavelengths; the values of N are in good agreement with the reference data in the literature [1].

Looking at the calculated curves in Fig. 6.1, we can see that Ψ is very well reproduced, while the predicted trend for Δ does not agree with the experiments; moreover, the roughness layer is completely rejected by the best-fit procedure. Such a behaviour indicates that our representation of the substrate is slightly over-simplified, and suggests the possible presence of a surface layer, with optical constants slightly different from the substrate, whose presence does not change the intensity of the reflected fields (related to Ψ), but modifies their phases (hence Δ). This is a reasonable assumption, especially since the hygroscopicity of LiF might favour the adsorption of water molecules; furthermore, the polishing procedure could have left some residuals contaminating the surface, or have induced some morphological disorder or defects like colour centers. Accounting for these contributions requires a more detailed inspection of the surfaces, which is far beyond the scope of the current analysis. Nevertheless, fitting Ψ already provides accurate optical constants for the substrate, so any further improvement of the model is not strictly necessary.

6.1.2 Nanostructured LiF Templates

We proceed by considering the optical properties of the substrates following the further deposition of LiF. In Sect. 4.1 we saw that elongated ripples spontaneously form after the deposition of ≈ 250 nm of LiF. The Ψ and Δ spectra corresponding to this situation, reported in Figs. 5.2 and 5.3, show, as expected, several differences with respect to the bare crystals. Ψ and Δ are both characterized by evident oscillations, and they acquire a slight but clear anisotropy when the plane of incidence is oriented parallel or perpendicular the ripples. Such an optical birefringence can be directly related to the intrinsic morphological anisotropy of the ripple structure, while the oscillations are most likely due to the interference between multiple reflections inside the deposited film (see Sect. 2.1.3). In particular, the latter observation implies that the molecular-beam-deposited layer of LiF has a refractive index different from the bulk crystal, probably due to a not perfect recombination of the Li and F ions during the desublimation of the gaseous phase, and by the consequent proliferation of structural defects; some contaminants can also get adsorbed and incorporated, since the typical base pressure during deposition is in the 10^{-7} mbar range.

We can exploit the optical interference to finely estimate the thickness of deposited LiF film. This can be easily seen for the ideal situation (sketch in Fig. 6.3) in which the system is composed of a flat homogeneous thin film (N_1) in air (N_0), supported on a semi-infinite homogeneous substrate (N_2); for simplicity we assume the refractive indices N_i as real constants (and $N_0 = 1$). Referring to Fig. 6.3, the constructive interference occurs whenever the difference between the optical paths travelled by the reflected beams is an integer multiple m of the wavelength, i.e. [2]

$$N_1(\overline{BC} + \overline{CD}) - N_0\overline{AD} = 2dN_1\cos\theta_1 = m\lambda_m = m\frac{hc}{E_m}$$

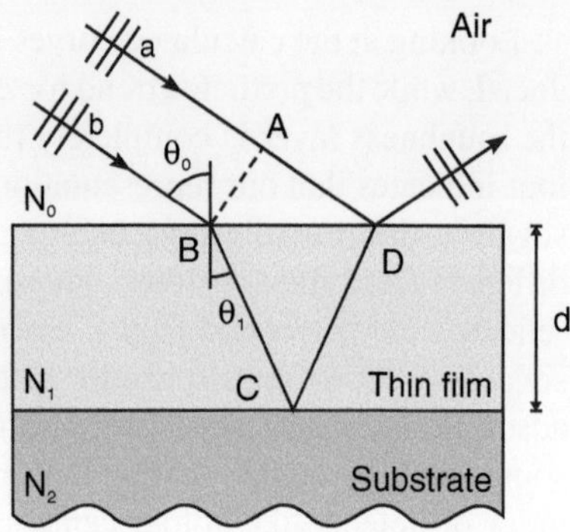

Fig. 6.3 Schematic representation for the reflection of an EM plane wave from a thin film on a substrate. The rays a and b are in phase in A and B, and sum up in D, respectively after a direct reflection from the surface and a first reflection from inside the film

where $E(\lambda) = hc/\lambda$ is the photon energy. The period of the interference figure (for example as a function of E) is then given by

$$\Delta E = E_{m+1} - E_m = \frac{hc}{2dN_1 \cos\theta_1},$$

directly related to the thickness and refractive index of the film. In this simple example, the period ΔE is independent of the energy and inversely proportional to d; in all real cases, however, this is not true due to the dispersion of the refractive indices, so a complete analysis of the optical constants is required to estimate the thickness.

In order to characterize the nanostructured substrates, we proceed as in the previous case, building an optical model and then performing a best-fit procedure to extract the optical constants. We extend the model employed for the bare LiF(110), including a second Cauchy layer to describe the deposited layer, that requires optical constants slightly different from the substrate. Then, in order to reproduce the anisotropy of Ψ and Δ, we add on top of the "film" layer a birefringent surface layer, having different refractive indices in the directions parallel and perpendicular the ripples, which accounts for the rippled surface morphology. An accurate treatment would require to fully consider the sawtooth-like shape of the ripples; here, instead, we apply a simpler approximation, treating the ripples as parallel cylinders aligned in the plane of the sample. This is implemented by combining two Bruggeman effective medium layers, with depolarization factors $L_{\parallel} = 0$ along the ripples and $L_{\perp} = 0.5$ in the perpendicular directions (see Fig. 2.11c), in a birefringent material placed on top of the film; both the EMA layers are composed in equal proportion of air and of the underlying film material. As in the previous case, we neglect any effect of surface contamination or disorder. The overall structure of the model is sketched in Fig. 6.4c. All the components employed (EMAs, birefringent layers, etc.) are standard elements provided by the WVASE software.

The best-fit Ψ and Δ curves corresponding to the experimental spectra of a nanopatterned LiF (Fig. 5.3) are reported in Fig. 6.4a, b. The best-fit thickness of the film and of the roughness layer, for the particular sample under scrutiny, were 249 and 4.2 nm, respectively, while a ripple periodicity of about 35 nm was deduced

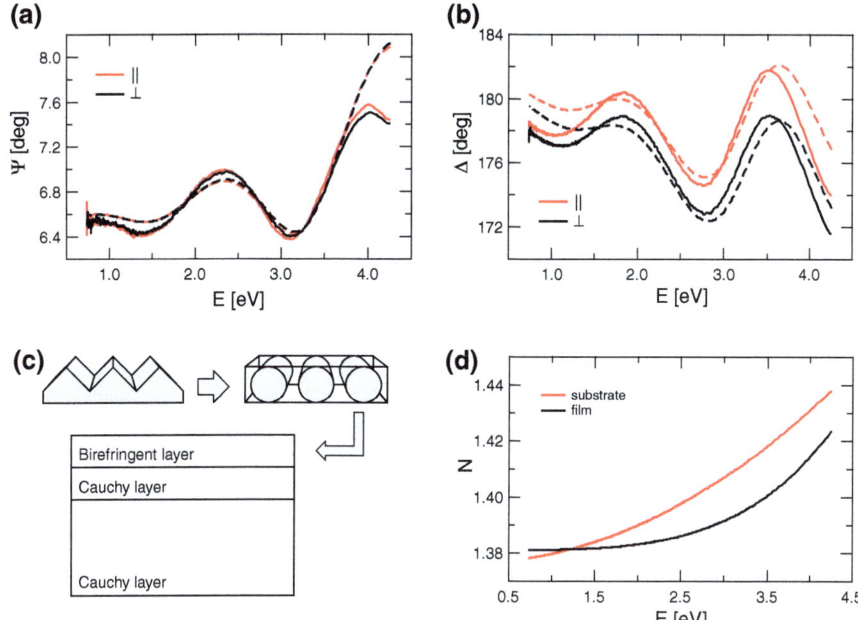

Fig. 6.4 Panels **a**, **b** experimental (*continuous lines*) and calculated (*dashed lines*) Ψ and Δ curves for LiF substrates after homoepitaxial deposition of 250 nm of LiF, at incidence $\theta = 50°$ and plane of incidence either parallel (*red lines*) and perpendicular (*black lines*) the ripples ridges. Panel **c** sketch of the optical model employed for the calculations. Panel **d** real parts of the calculated refractive indices for the substrate (*red line*) and the deposited layer of LiF (*black line*)

from AFM. The agreement between the simulations and the experiments is again not excellent, but the main features of the ellipsometric spectra are all well reproduced. The calculated curves for the plane of incidence parallel and perpendicular the ripples direction are indeed slightly shifted, the differences being more pronounced in Δ than in Ψ, and in agreement with the experiments, indicating that the approximation to cylinder is good for a first order analysis. The thickness of the effective layer was instead quite underestimated in comparison to the AFM data; in the ideal case of a perfect ripple structure, the depth of the valleys is equal to half of the periodicity, corresponding to ≈17 nm for the current case; if we further include some disorder and roughness, then we would expect a thickness greater than ≈20 nm, which is clearly in contrast with the value of merely 4.2 nm extracted from the fit.

The calculated real parts of the refractive indices for the substrate and the film are shown in Fig. 6.4d (the imaginary parts are zero at all wavelengths). The two curves are slightly different, with variations within the 1 % of the absolute values, leading to the observed interference effects; in particular, moving towards higher energies, the real refractive index of the film increases more rapidly than the substrate, indicating the possible presence of absorptions at energies lower than the band gap, consistent with the formation of crystal defects or color centers [3, 4] during the deposition of LiF.

Summarizing the results of this section, we characterized the optical properties of the LiF substrates before and after the deposition of LiF films, performed for inducing the formation of the ripples nanostructures. The refractive index of bulk LiF was calculated for the bare LiF(110) samples, and found to be isotropic within the experimental accuracy. The deposited layer of LiF, instead, exhibited a slightly different behaviour with respect to the bulk, and showed a weak optical anisotropy referable to the ripple morphology. The variations of the refractive indices were found to lie within the 1 % of the bulk value, and the birefringence was much weaker than the optical anisotropy observed on the gold NPs arrays. Therefore, in the framework of the model presented in the next section, we will neglect these effects, and treat the substrates as homogeneous and isotropic materials.

6.2 2-Dimensional Arrays of Gold Nanoparticles

We now focus on the optical modelling of the 2D arrays of gold NPs. The NP in such arrays have non-spherical shape and are coherently aligned in either square or rectangular lattices. The LSP characteristics of the arrays depend both on intrinsic effect, like the single-NP shape, introducing anisotropic NP polarizability, and on the arrangement of the NPs in the arrays, possibly inducing anisotropic EM coupling between the NPs. In order to separately assess the role of each effect, the experiments must be backed by appropriate model calculations of the LSP, that keep intrinsic and collective effects in due account.

In general, due to the complexity of the involved systems, numeric methods are the most suitable to efficiently solve the optical properties of non-spherical particles; these include, for example, discrete dipole approximation [5, 6], T-matrix [7, 8] and finite elements [9] models. Here, however, we approach the optical problem applying a simpler mean field theory, based on effective medium approximations. Although this approach is less rigorous than the aforementioned methods, it is much easier to implement, it can include the main effects of the particles shape and of the presence of mutual interactions, and handles the presence of morphological disorder in a relatively straightforward manner.

6.2.1 Model

The 2D arrays of gold NPs are quite elaborated systems, presenting optically and morphologically anisotropic structures, facets tilted from the sample plane, and nearly-contacting metallic particles with complex truncated shapes. In order to describe such systems, without resorting to unnecessary sophisticated computational methods, some simplifications are therefore necessary. The major approximations we performed involve the following parameters:

- NP arrangement, Fig. 6.5a. We consider the ideal case where the substrates have a perfect ripple morphology; the NPs chains are then regularly spaced across the ripples, with the same periodicity Λ of the substrate. Furthermore, following the observation that the particles are partially aligned also between adjacent chains (see Fig. 4.9), we also assume the NPs spacing along the chains to be well defined, so that the particles can be schematized arranged on a 2D rectangular lattice.
- NP shape, Fig. 6.5b. The gold NPs are supported on the facets of the ripples, thus presenting a truncated shape tilted with respect to the plane of the sample. We simplify this geometry by considering non-truncated and non-tilted ellipsoids, having their principal axes either parallel or normal to the surface plane. This is not a drastic approximation, because the annealing procedure is expected to smooth the NPs surface, while the tendency of gold to dewet suggests a low contact area between the NPs and LiF.
- Layers, Fig. 6.5c. The surface region of the Au/LiF nanostructures is complex. It is composed of several materials (LiF, gold, air), separated by interfaces with complex geometries (ripples, nanoparticles). In order to include in the model the contributions of the most significant parameters, without complicating the model, we make extensive use of effective medium approximations (Sect. 2.2). Considering the previous characterization of the LiF templates, we decompose the system in two main parts, a substrate and a thin film. The substrate accounts for the optical contribution of the bulk LiF crystal and the deposited LiF; since we found variations of the refractive indices within the 1 % of the absolute values, here we assume a homogeneous and isotropic material, with optical constants reported in Fig. 6.2. The "film", instead, represents the surface layer accomodating the nanoparticles. In the previous section we treated the ripples as cylinders, now, however, as the ripple birefringence is much lower than the optical anisotropy of the nanoparticles, we apply a much simpler approximation, describing the ripples as a homogeneous effective host, with dielectric constant that are a linear combination of the optical constants of LiF and air. We also employ effective medium approximation for the Au nanostructures, calculating an anisotropic effective dielectric tensor ε_{eff} for the "film", which includes the contributions of the NPs shape and mutual EM coupling [see Eq. (6.14)].

Under these assumptions, we model the Au/LiF nanostructures as an ensemble of $N \gg 1$ identical and aligned polarizable inclusions, immersed in a homogeneous and isotropic host and placed on top of a homogeneous and isotropic substrate, as sketched in Fig. 6.5b. We set the coordinate system with the axis z normal the substrate and the axes x and y on the surface, the axis y being parallel the ripples direction. All the inclusions have the same ellipsoidal shape, with principal semiaxes a_x, a_y and a_z along the cartesian axes, and are arranged on a rectangular grid with first-neighbour spacings d_x and d_y; the center of the inclusions is placed at distance $d_z > a_z$ from the substrate/film interface. The substrate, the host medium and the inclusions have, respectively, dielectric functions $\varepsilon_s(\omega)$, $\varepsilon_h(\omega)$ (both purely real) and

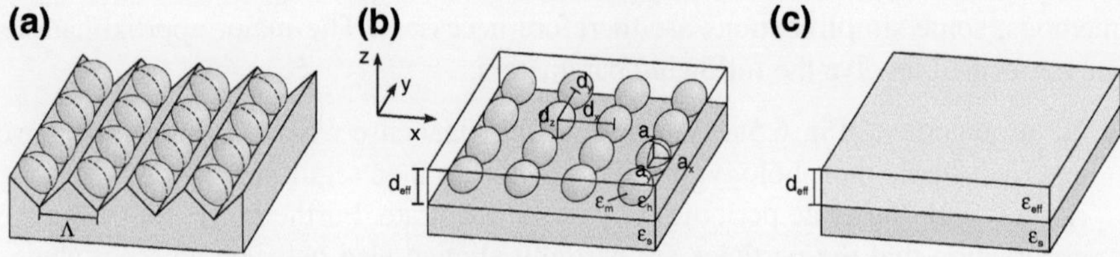

Fig. 6.5 Schematic representation of the approximations applied to the gold NPs. Panel **a** gold NPs supported on a perfect and regular LiF ripple structure, and arranged on a rectangular mesh. Panel **b** the NPs are assumed ellipsoidal shaped, with axes aligned to the plane of the sample, and immersed in a homogeneous and isotropic host. Panel **c** the host and the NPs are replaced by a homogeneous effective medium with anisotropic dielectric constant $\boldsymbol{\varepsilon}_{eff}$

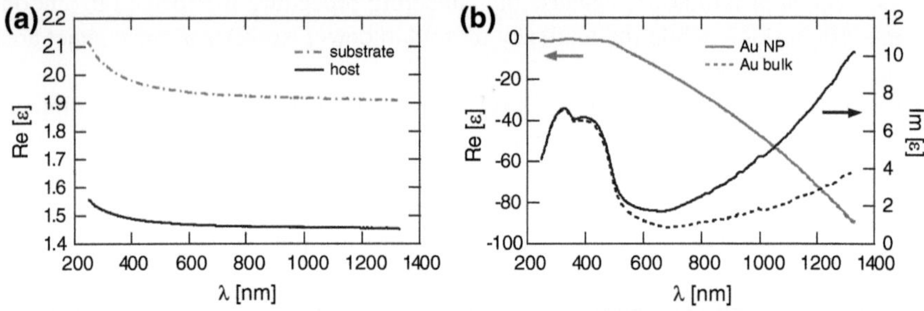

Fig. 6.6 Panel **a** real parts of the dielectric constants ε_s for the substrate (*red line*) and ε_h for the host (*black line*) as employed in the calculations; the corresponding imaginary parts are null at all wavelengths. Panel **b** real and imaginary parts of the dielectric constant ε_m for the metallic inclusions (*continuous lines*), compared to the corresponding values for bulk gold (*dashed lines*)

$\varepsilon_m(\omega)$ (complex), reported in the graphs in Fig. 6.6. ε_s is converted from the refractive index of bulk LiF (Fig. 6.2), while ε_h is the average between ε_s and vacuum ($\varepsilon = 1$); ε_m is instead obtained by extracting the dielectric constant of gold from ellipsometric measurements performed on clean bulk samples (Au(111)/Mica from PHASIS, inset of Fig. 5.4), and then applying the finite size corrections according to Eq. (2.57).

In the quasi-static approximation (Sect. 2.3.1) the dipolar polarizability tensor $\boldsymbol{\alpha}$ of the individual inclusions is diagonal and its principal components are given by Eqs. (2.60) and (2.42). In order to account for the morphological disorder of the nanoparticles, the size/shape dispersion unavoidably encountered in our samples are effectively converted into a corresponding distribution of depolarization factors L_γ, and accordingly replace Eq. (2.60) with [10]:

$$\alpha_\gamma = \int_0^1 \mathcal{P}(L_\gamma)\, \alpha_\gamma^{\mathrm{MLWA}}\, dL_\gamma \tag{6.1}$$

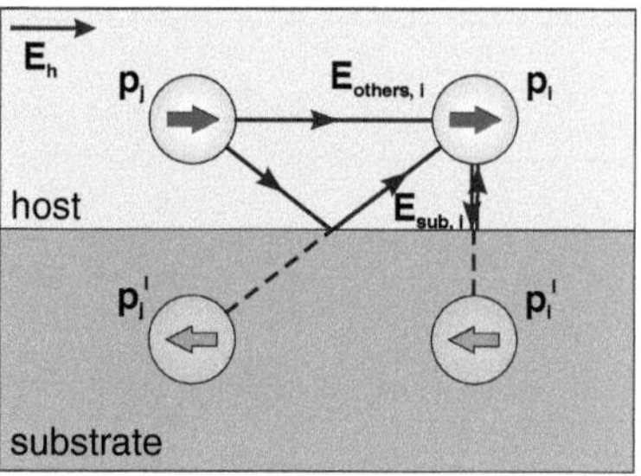

Fig. 6.7 Sketch of the dipolar contributions to the local electric field acting on each particle, and representation of the image dipoles induced inside the substrate

where $\mathcal{P}(\eta)$ can be any arbitrary distribution and the mean values for L_γ are chosen according to (2.43).

The system is excited by an external electric field $\mathbf{E}_{ex}(\mathbf{r}, t) = \mathbf{E}_{ex}e^{i(\omega t - \mathbf{q}\cdot\mathbf{r})}$, with frequency ω and wave vector $\mathbf{q}$; in the quasi-static limit the electric field inside each inclusion is uniform, thus $q \gg 1/a_i$. We suppose the incident field to be parallel to one of the principal axes ξ of the ellipsoids (ξ can be either x, y or z), $\mathbf{E}_{ex} = E_{ex}\hat{\xi}$. Omitting the time factors $e^{i\omega t}$, the dipolar moment $\mathbf{p}_i$ induced inside the i-th inclusion, with center located at $\mathbf{r}_i$, is proportional to the local electric field $\mathbf{E}_{\mathrm{loc},i}$ acting at $\mathbf{r}_i$ [see Eq. (2.41)]

$$\mathbf{p}_i = \varepsilon_0 \boldsymbol{\alpha} \otimes \mathbf{E}_{\mathrm{loc},i} \tag{6.2}$$

The local field $\mathbf{E}_{\mathrm{loc},i}$ is the sum of several contributions, as sketched in Fig. 6.7:

$$\mathbf{E}_{\mathrm{loc},i} = \mathbf{E}_{h,i} + \mathbf{E}_{\mathrm{others},i} + \mathbf{E}_{\mathrm{sub},i} \tag{6.3}$$

- $\mathbf{E}_{h,i} = \varepsilon_h \mathbf{E}_{ex}e^{-i\mathbf{q}_h\cdot\mathbf{r}_i}$: the exciting field propagating inside the host, in the absence of any inclusion; it differs from $\mathbf{E}_{ex,i}$ due to the polarization of the host;
- $\mathbf{E}_{\mathrm{others},i}$: the sum of the dipolar fields generated at the position $\mathbf{r}_i$ by all the other inclusions;
- $\mathbf{E}_{\mathrm{sub},i}$: the sum of the dipolar radiation generated by the inclusions and reflected from the substrate.

In first approximation $\mathbf{E}_{\mathrm{sub},i}$ can be calculated within the image charge model, so that the contribution of each dipole $\mathbf{p}_j$ is equivalent to the field generated by a dipole $\mathbf{p}'_j$ situated at specular position with respect to the interface and given by [11, 12]

$$\mathbf{p}'_j = \frac{\varepsilon_s - \varepsilon_h}{\varepsilon_s + \varepsilon_h}\begin{bmatrix} -1 & 0 & 0 \\ 0 & -1 & 0 \\ 0 & 0 & 1 \end{bmatrix} \otimes \mathbf{p}_j = \mathrm{M} \otimes \mathbf{p}_j, \tag{6.4}$$

The dipolar fields generated by $\mathbf{p}_j$ and $\mathbf{p}'_j$ at $\mathbf{r}_i$ are expressed according to (2.48), which we rewrite using the dipolar interaction tensors t_{ij} and t'_{ij}:

$$\mathbf{E}_j(\mathbf{r}_i) = t_{ij} \otimes \mathbf{p}_j$$
$$\mathbf{E}_j^I(\mathbf{r}_i) = t_{ij}^I \otimes \mathbf{p}_j^I$$

Then, Eq. (6.2) becomes

$$\mathbf{p}_i = \varepsilon_0 \alpha \otimes \left(\mathbf{E}_h e^{-i\mathbf{q}_h \cdot \mathbf{r}_i} + \sum_{j \neq i} t_{ij} \otimes \mathbf{p}_j + \sum_j t_{ij}^I \otimes \mathbf{p}_j^I \right) \tag{6.5}$$

Introducing

$$s_{ij} = (1 - \delta_{ij})t_{ij} + t_{ij}^I \mathrm{M} \tag{6.6}$$

we can rewrite (6.5) in the more compact form

$$\mathbf{p}_i = \varepsilon_0 \alpha \otimes \left(\mathbf{E}_h e^{-i\mathbf{q}_h \cdot \mathbf{r}_i} + \sum_j s_{ij} \otimes \mathbf{p}_j \right) \tag{6.7}$$

The calculation of the sum in (6.7) is greatly simplified for inclusions disposed on a rectangular lattice with principal axes aligned the lattice vectors. In fact, the dipoles $\mathbf{p}_i$ are parallel to the exciting electric field, and the only non-zero component is thus the one lying along ξ, which reads

$$p_i^\xi = \varepsilon_0 \alpha^\xi \left(E_h e^{-i\mathbf{q}_h \cdot \mathbf{r}_i} + \sum_j s_{ij}^{\xi\xi} p_j^\xi \right) \tag{6.8}$$

Neglecting boundary effects, all the induced dipoles have the same magnitude p, while the relative phases vary due to the spatial oscillations of the electric field; if we substitute $p_j \equiv p \, e^{-i\mathbf{q}_h \cdot \mathbf{r}_j}$ in (6.8) and solve for p, we obtain

$$p^\xi = \varepsilon_0 \frac{\alpha^\xi}{1 - \varepsilon_0 \alpha^\xi S^{\xi\xi}} E_h = \varepsilon_0 \varepsilon_h \frac{\alpha^\xi}{1 - \varepsilon_0 \alpha^\xi S^{\xi\xi}} E_{ex} \tag{6.9}$$

where $S^{\xi\xi} \equiv \sum_j s_{ij}^{\xi\xi} e^{-i\mathbf{q} \cdot \mathbf{r}_{ij}}$ and $\mathbf{r}_{ij} = \mathbf{r}_j - \mathbf{r}_i$.

We can now introduce the *effective polarizability tensor* A, which expresses the polarizability of the metallic inclusions accounting for their reciprocal dipolar interactions:

$$A^{\gamma\gamma} = \frac{\alpha^\gamma}{1 - \varepsilon_0 \alpha^\gamma S^{\gamma\gamma}}, \quad \gamma = x, y, z \tag{6.10}$$

Then, the induced dipoles can be finally written as

$$\mathbf{p} = \varepsilon_0 \varepsilon_h A \otimes \mathbf{E}_{ex} \tag{6.11}$$

We can now formulate an effective medium approximation for the metallic inclusions. Given the isolated nature of the inclusions, we follow an approach similar to the Maxwell-Garnett EMA (Sect. 2.2), modified in order to account the ordered arrangement of the inclusions and the presence of the substrate (cfr. also refs. [13–15]).

We note that the induced dipoles Eq. (6.11) within the array have the same expression (2.41) for a single isolated particle, except for the polarizability tensor $\boldsymbol{\alpha}$ replaced by $\boldsymbol{A}$.

Then, we can treat the inclusions as isolated, and write the total polarization density as

$$\mathbf{P} = n\mathbf{p} = \varepsilon_0 \varepsilon_h n \boldsymbol{A} \otimes \mathbf{E}_{ex}, \tag{6.12}$$

where $n = 1/d_x d_y d_{\text{eff}}$ is the number of inclusions per unit volume, and d_{eff} the thickness of the host. The effective dielectric tensor $\boldsymbol{\varepsilon}_{\text{eff}}$ for the inclusions can be evaluated from the definition of the electric displacement field $\mathbf{D}$

$$\mathbf{D} = \varepsilon_0 \boldsymbol{\varepsilon}_{\text{eff}} \otimes \mathbf{E}_{ex} = \varepsilon_0 \varepsilon_h \mathbf{E}_{ex} + \mathbf{P} = \varepsilon_0 \varepsilon_h \mathbf{E}_{ex} + \varepsilon_0 \varepsilon_h n \boldsymbol{A} \otimes \mathbf{E}_{ex} \tag{6.13}$$

where we can see that $\boldsymbol{\varepsilon}_{\text{eff}}$ is diagonal similar to $\boldsymbol{A}$. Solving separately for the three principal components we finally obtain

$$\frac{\varepsilon_{\text{eff}}^{\xi\xi}}{\varepsilon_h} = 1 + nA^{\xi\xi}, \quad \xi = x, y, z \tag{6.14}$$

Using the dielectric tensor $\boldsymbol{\varepsilon}_{\text{eff}}$ we can represent the inclusions and the host as a homogeneous and biaxially anisotropic film (Fig. 6.5c), still preserving all the informations on the NPs shapes and interactions. The optical response of the whole system can then be easily calculated using the Fresnel coefficients (2.26) for the thin-film/substrate model. In this respect, however, the refractive indices for the s and p components of the electric field are required, whereas the dielectric tensor $\boldsymbol{\varepsilon}_{\text{eff}}$ is expressed in xyz coordinates. We need therefore to rewrite the components of $\boldsymbol{\varepsilon}_{\text{eff}}$ in sp coordinates.

In general the reflection of light from a biaxially anisotropic film is a quite complex phenomenon, especially when the principal axes of the dielectric tensor are not parallel to the interfaces or the plane of incidence [16, 17]. In this case, however, we just need to replicate the geometries applied in the experiments, i.e. plane of incidence either parallel (y axis) or perpendicular (x axis) the ripples.

Therefore, we suppose the plane of incidence parallel to the xz plane, as in Fig. 6.8, corresponding to the perpendicular ($\perp$) geometry; the calculations in parallel ($\parallel$) geometry are exactly the same, except swapping the axes x and y.

We consider an EM plane wave travelling in the xz plane and approaching the surface at an angle of incidence θ. When the electric field is s-polarized, it always oscillates along the y direction, independent of θ, so the wave propagates according to solely the dielectric function $\varepsilon_{\text{eff}}^{yy}$. The (complex) refractive index $\tilde{n}_{\text{eff}}^{s}$ is then simply

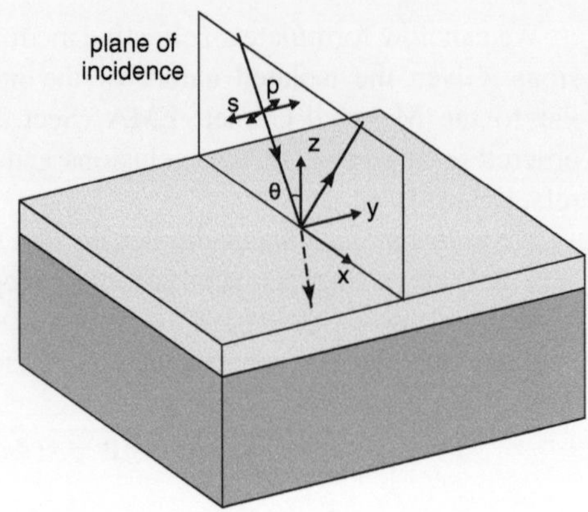

Fig. 6.8 Sketch of the reflection geometry, highlighting the differences between the sp coordinate systems, employed for the Fresnel coefficients (2.23), and the xyz principal axes of the effective dielectric tensor $\boldsymbol{\varepsilon}_{\text{eff}}$

given by

$$\tilde{n}^{s}_{\text{eff}} = \sqrt{\varepsilon^{yy}_{\text{eff}}} \tag{6.15a}$$

On the contrary, for p-polarized light the electric field lies in the xz plane, therefore the p wave "feels" the combined effects of the different dielectric functions $\varepsilon^{xx}_{\text{eff}}$ and $\varepsilon^{zz}_{\text{eff}}$, with relative weights determined by the angle θ. $\tilde{n}^{p}_{\text{eff}}$ can be deduced from the *refractive index ellipsoid* and writes [16–18]

$$\tilde{n}^{p}_{\text{eff}} = \tilde{n}^{p}_{\text{eff}}(\theta) = \sqrt{\frac{\varepsilon^{xx}_{\text{eff}}\, \varepsilon^{zz}_{\text{eff}} + (\varepsilon^{zz}_{\text{eff}} - \varepsilon^{xx}_{\text{eff}})\sin^2\theta}{\varepsilon^{zz}_{\text{eff}}}} \tag{6.15b}$$

The reflection and transmission of light can now be computed at any incidence angle and state of polarization of incoming light by applying Eq.(2.6) to the configuration of Fig. 6.5c, and using $\tilde{n}^{p,s}_{\text{eff}}$ for the film and $\tilde{n}_s = \sqrt{\varepsilon_s}$ for the substrate.

6.2.2 Optical Anisotropy of Self-Organized Au Nanoparticles Arrays

With a theoretical description framework available, we can now successfully address the plasmonic properties of the arrays of gold NPs and separately investigate the contributions of the NPs shape and of the mutual interactions to the collective optical anisotropy of the systems. In Sect. 5.2.3 we already discussed qualitatively the influence of the NPs aspect ratio and spacings; here, we apply the model developed in Sect. 6.2.1 to quantitatively replicate the optical response of the arrays and estimate the effects of the morphological characteristics on the LSP resonances.

The calculations are performed employing the morphological parameters deduced by AFM analysis as geometrical input parameters for the model, in order to prevent the achievement of incidental agreement between data and model with non realistic sample characteristics.

The first set of morphological constraints that we apply concerns the thickness of the effective medium layer and the positioning of the inclusions. Assuming an ideal ridge-and-valley structure, with regularly spaced facets tilted of $45°$ with respect to the substrate, we set the thickness of the effective layer d_{eff} to half the ripples periodicity Λ ($d_{\mathrm{eff}} = \Lambda/2$), and we assume the particles centered at half height inside the layer ($d_z = d_{\mathrm{eff}}/2 = \Lambda/4$).

The in-plane periodicities of the inclusions, d_x and d_y ($d_y = \Lambda$), are chosen according to the mean values extracted from the AFM data, neglecting in first approximation the contributions of their finite spread on the LSP response. In fact, while there are ample demonstrations that the LSP resonances in isolated NPs pairs are a strong function of their mutual separation [19–22] (see also Sect. 2.3.4), the 2D rectangular symmetry of our arrays has the positive effect of limiting the impact of the interparticle dipolar coupling on the LSPs, via a partial cancellation of the dipole interactions between the NPs (Fig. 6.16). This effect, discussed more in detail in subsection "Discussion", is qualitatively supported by the observation of remarkably narrow LSP peaks, compared to what would be expected considering the spread in the NP-NP distance [19–22]. Moreover, such cancellation effects also makes the LSP response less sensitive to the morphological disorder.

Considering instead the geometry of the nanoparticles, the LSP can be critically affected by the dispersion of sizes and shapes. In our samples, the NP dimensions are such that multipolar LSP are fully negligible, therefore the presence of a distribution of NP sizes affects the LSP resonances mainly via the appearance of intrinsic finite-size corrections to the bulk Au dielectric constant [23–25] (cfr. Sects. 2.3, 2.3.2). These effects have been included in the model, as already explained before (see Fig. 6.6b), although, for the typical mean size and deviation under consideration here, they yield a relatively minor impact on the LSP [26]. NP shape effects, namely the distribution of in-plane aspect ratios, provide instead a significant LSP broadening and weakening mechanisms, through the onset of a corresponding distribution of depolarization factors [Eq. (6.1)].

As described in Sect. 4.2, we can independently control the NPs aspect ratio and spacing by appropriately tuning, during the samples fabrication, the substrate temperature and the thickness of the deposited gold layer. We exploit this possibility by realizing *two different classes of samples*, respectively featuring in-plane isotropic Au NPs arranged on a rectangular lattice and in-plane elongated Au NPs arranged on a square grid. In both cases, an optical anisotropy arises in the system, respectively originating from the anisotropic EM coupling in the array and from the single-NP intrinsic anisotropic polarizability, which in turn determines anisotropic EM interactions between the NPs.

In the next paragraphs we will review the morphological and optical characterizations for such classes of samples, and then exploit the capabilities of our model to assess the different contributions of the intrinsic and collective effects on the LSP.

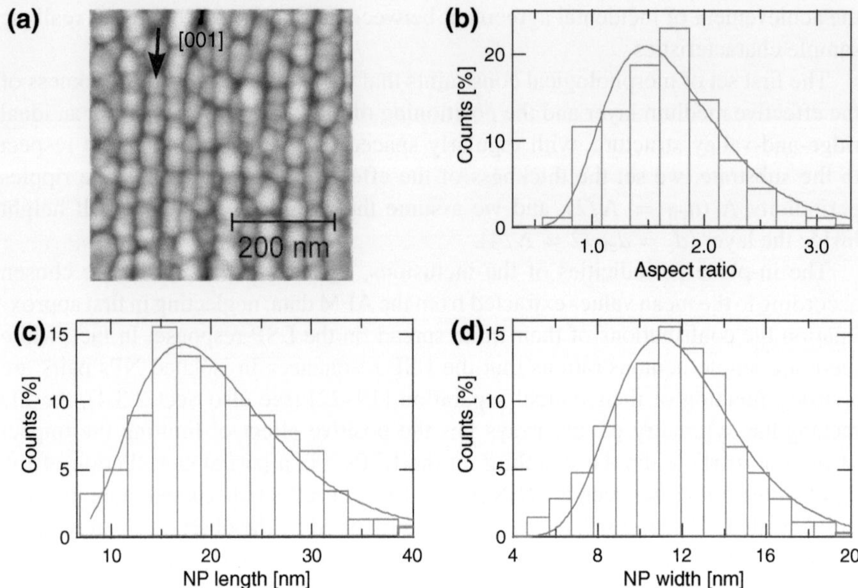

Fig. 6.9 Panel **a** AFM image of a nanopatterned LiF(110) sample with $\Lambda \approx 40$ nm, following the grazing deposition of ≈ 5 nm of gold at $T = 100\,°C$ and annealing at $T = 400\,°C$ ("square" configuration). Panels **b, c, d** statistical distributions of the NP in-plane aspect ratio and semiaxes along and across the LiF ridges, respectively. The continuous lines are best-fit lognormal probability density functions. See text for details

Morphological Characterization

The quantitative morphological parameters corresponding to the first sample are summarized in Fig. 6.9 (cfr. the AFM measurement of Fig. 4.8c). In panel (a) we report a representative AFM image of the LiF(110) surface following homoepitaxial growth of LiF at $T = 350\,°C$, deposition of $t_{Au} \approx 5$ nm of Au and annealing at $T = 400\,°C$. The ripple morphology has a mean periodicity $\Lambda \approx 40$ nm, and the formation of Au NPs slightly elongated along the ridges direction after the annealing can be easily noticed.

The experimental distribution of the NPs sizes were obtained by careful analysis of the AFM data. First, the NPs present in the images were isolated by the application of a threshold setting to separate them from the background. The NPs isolated by this procedure in a $1\,\mu m^2$ AFM image, containing about one thousand of particles, had their perimeters fitted with sets of coherently aligned ellipses. The ellipses axes are then used to provide the in-plane ellipsoids dimensions and the corresponding aspect ratios. Due to AFM tip-convolution effects, the heights of the ellipsoids cannot be precisely assessed, thus leaving the "normal" ellipsoid axis the only relatively free parameter of the model.

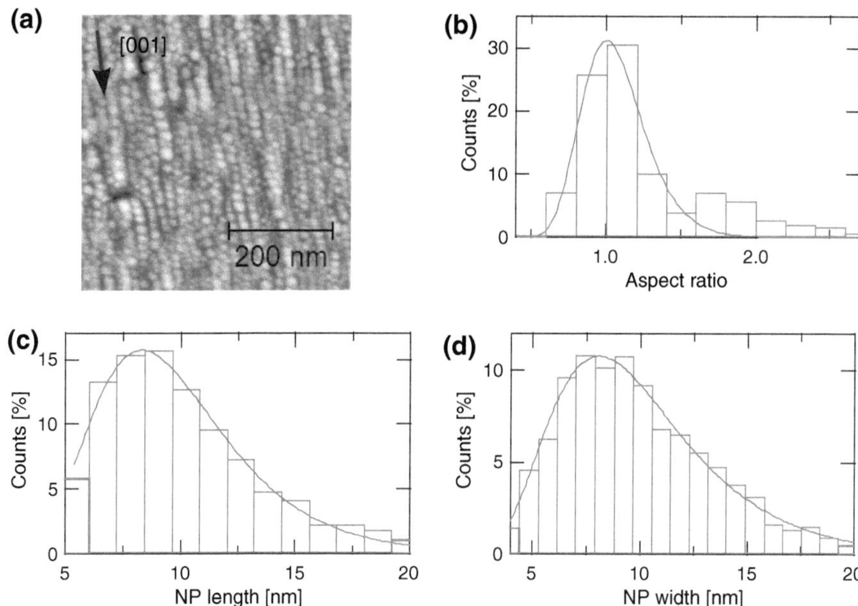

Fig. 6.10 Panel **a** AFM image of a nanopatterned LiF(110) sample with $\Lambda \approx 30$ nm, following the grazing deposition of ≈ 3 nm of gold at $T = 100\,^{\circ}$C and annealing at $T = 400\,^{\circ}$C ("rectangular" configuration). Panels **b, c, d** statistical distributions of the NP in-plane aspect ratio and semiaxes along and across the LiF ridges, respectively. The *continuous lines* are best-fit lognormal probability density functions. See text for details

The distributions of the in-plane NP semiaxes, and the corresponding aspect ratio spread, deduced from the image in Fig. 6.9a, are reported in the graphs in Fig. 6.9b–d. The semiaxes of the NPs were found to obey a lognormal distribution, with mean and standard deviation $a_y = (17 \pm 7)$ nm along the LiF ridges and $a_x = (11 \pm 3)$ nm across the ridges; the in-plane aspect ratio also exhibited lognormal distribution, with mean and standard deviation of 1.5 ± 0.5. A similar statistical treatment of the images also allowed to deduce the mean periodicities of the particles along the LiF ridges and across the NPs chains, respectively reading $d_y = (41 \pm 5)$ nm and $d_x = (40 \pm 5)$ nm, respectively. The close similarity of these two values suggests that the NPs can be schematically modelled as lying on a square mesh. We will then call this class of samples as the "square" arrays, referring to the NPs arrangement.

In Fig. 6.10 we report a representative set of data recorded for the second class of samples under scrutiny. The AFM image in panel (a) shows the morphology of a nanopatterned substrate characterized by a mean periodicity $\Lambda \approx 30$ nm (obtained by LiF homoepitaxy at $T = 300\,^{\circ}$C) after the deposition of a gold layer with $t_{Au} \approx 3$ nm thickness and subsequent annealing at $T = 400\,^{\circ}$C. Under these fabrication conditions the nanoparticles grow smaller compared to the "square" case, and exhibit a roughly circular in-plane aspect. Applying the same statistical analysis to the AFM data as for the previous case, the NP mean size, aspect ratio and arrangement can be quantitatively assessed. In this case, we found that the size distributions for the in-plane semiaxes, in the directions along and across the LiF ridges, were characterized by very similar parameters, exhibiting a mean and a standard deviation of $a_x = a_y = (8.5 \pm 3.0)\,$nm, corresponding to an aspect ratio of 1.0 ± 0.3. Mean NP spacings of $d_y = (20 \pm 5)\,$nm along the chains and $d_x = (30 \pm 5)\,$nm across the ripples were found, indicating that the particles can be statistically thought as laying on a rectangular mesh. We will refer to this class of samples as the "rectangular" arrays.

Optical Characterization

In analogy with the previous cases, the optical response of the samples under scrutiny has been investigated by means of polarized light reflectivity, with the plane of incidence either along ($\parallel$) or across ($\perp$) the LiF ridges, in order to selectively discriminate the contributions of the individual L and T plasmonic modes (see Sect. 5.2.3).

In Figs. 6.11 and 6.12 we report, as open circles, R_S (panels (b)) and R_P (panels (c)) spectra measured for the "square" and the "rectangular" samples, respectively, at $\theta = 50°$ of incidence. For each polarization, the longitudinal and transverse LSP modes have been excited by fixing the plane of incidence in parallel or perpendicular configuration: L mode (black lines) excited in $R_S^{\perp}$, T mode (red lines) excited in $R_S^{\parallel}$, and vice versa for R_P.

The in-plane optical anisotropy of the system is particularly accentuated for the "square" configuration, reported in Fig. 6.11: looking at R_S, the L and T modes are excited by setting the plane of incidence perpendicular ($\perp$, Fig. 6.11c) and parallel ($\parallel$, Fig. 6.11b) to the LiF ridges, respectively, and are found at $\lambda_L^S = (597 \pm 3)\,$nm and $\lambda_T^S = (542 \pm 3)\,$nm, separated by $\approx 55\,$nm; the full widths at half maximum (Γ) of the L and T resonances are also markedly different, reading $\Gamma_L^S = (150 \pm 5)\,$nm and $\Gamma_T^S = (94 \pm 3)\,$nm.

In case of p-polarized light the EM excitation is orthogonal with respect to s polarization, therefore, for a given optical geometry ($\parallel$ or $\perp$), the LSP modes are exchanged: the L mode is observed in $\parallel$ geometry (Fig. 6.11a), while the T mode in $\perp$ geometry (Fig. 6.11d). The LSP characteristics obtained from R_P are $\lambda_L^P = (597 \pm 3)\,$nm and $\Gamma_L^P = (167 \pm 5)\,$nm for the L mode, $\lambda_T^P = (539 \pm 3)\,$nm and $\Gamma_T^P = (95 \pm 5)\,$nm for the T mode; the slight differences from the corresponding R_S values are due to the fact that in p polarization the electric field is not purely parallel the surface, but has also a normal component, which excites the LSP N mode.

The "rectangular" sample (Fig. 6.12) has instead a less pronounced optical anisotropy. The longitudinal LSP peak is found at $\lambda_L^S = (561 \pm 3)\,$nm and $\lambda_L^P = (567 \pm 3)\,$nm, red shifted by only 15-20 nm with respect to the transverse LSP located at $\lambda_T^S = \lambda_T^P = (546 \pm 3)\,$nm, and all the peaks exhibit roughly the same width of $\Gamma \approx 100\,$nm.

As a general consideration, we point out that the main difference between the R_S and R_P spectra is the overall intensity, as expected due to the different optical coefficients, while the position and the width of the resonances is preserved;

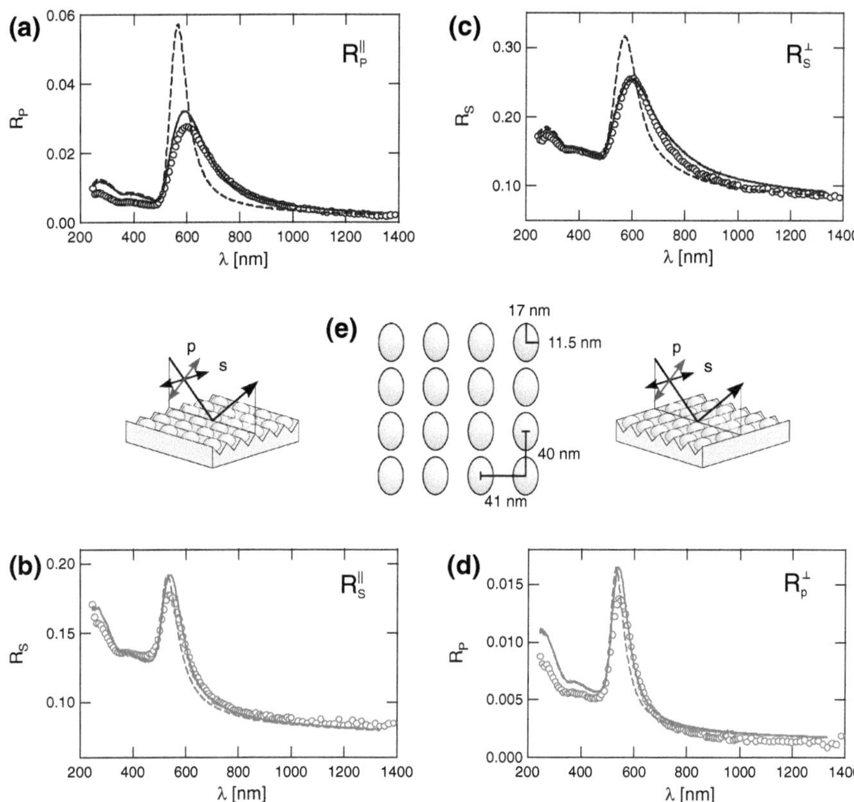

Fig. 6.11 "Square" sample. *Left panels* p-polarized (**a**) and s-polarized (**b**) reflectivity spectra measured with the plane of incidence parallel ($\parallel$) the LiF ridges. *Right panels* s-polarized (**c**) and p-polarized (**d**) reflectivity spectra measured with the plane of incidence perpendicular ($\perp$) the LiF ridges. *Red curves* (*bottom panels*): excitation of transverse LSP mode ($R_S^{\parallel}$, $R_P^{\perp}$); *black curves* (*top panels*): excitation of longitudinal LSP mode ($R_S^{\perp}$, $R_P^{\parallel}$). The continuous lines represent the corresponding reflectivities calculated according the model described in Sect. 6.2.1. The *dashed curves* were computed with the same morphological parameters like the previous ones but neglecting the dispersion of the depolarization factors. Panel **e** schematic diagram representing the NPs arrangement and shape employed in the optical calculations

in particular, no clear evidence of N LSP modes was found in any R_P curve, indicating that the polarizability for electric fields normal the surface plane is much lower than along the in-plane directions.

The corresponding reflectivities, computed according to the model for the gold nanoparticles, are plotted as continuous lines in the graphs of Figs. 6.11 and 6.12. The calculations were performed using the morphological parameters deduced from the AFM data, reported for reference in Table 6.1; the schematic diagrams of the NPs geometry and arrangement employed for the calculations are also reported in panels (a) of the same figures.

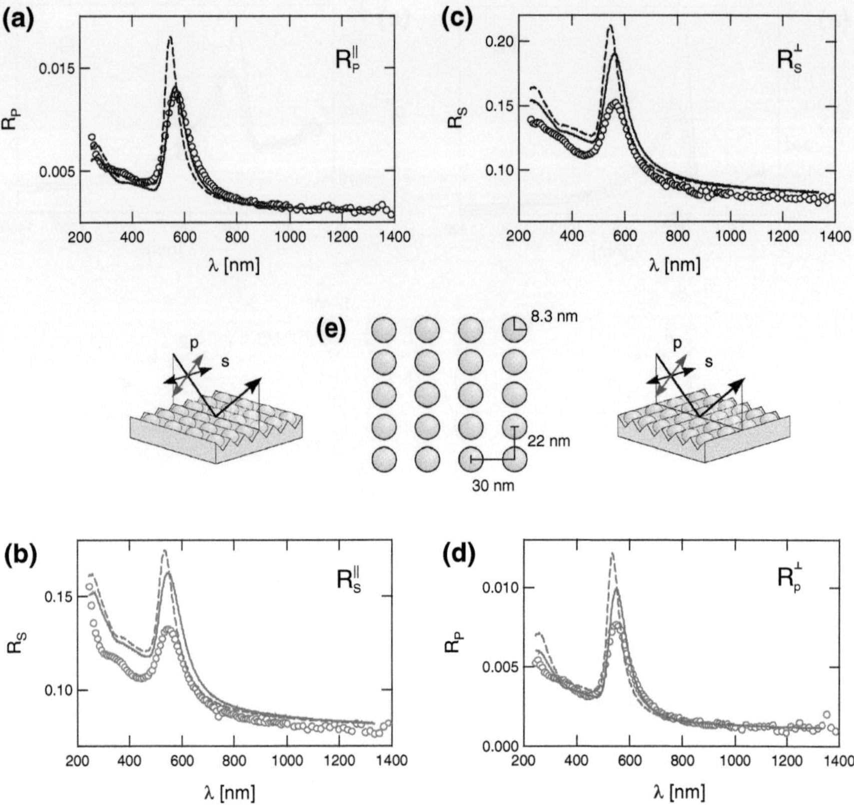

Fig. 6.12 "Rectangular" sample. *Left panels* p-polarized (**a**) and s-polarized (**b**) reflectivity spectra measured with the plane of incidence parallel ($\parallel$) the LiF ridges. *Right panels* s-polarized (**c**) and p-polarized (**d**) reflectivity spectra measured with the plane of incidence perpendicular ($\perp$) the LiF ridges. *Red curves* (*bottom panels*) excitation of transverse LSP mode ($R_S^\parallel$, $R_P^\perp$); *black curves* (*top panels*) excitation of longitudinal LSP mode ($R_S^\perp$, $R_P^\parallel$). The *continuous lines* represent the corresponding reflectivities calculated according the model described in Sect. 6.2.1. The *dashed curves* were computed with the same morphological parameters like the previous ones but neglecting the dispersion of the depolarization factors. Panel **e** schematic diagram representing the NPs arrangement and shape employed in the optical calculations

The distributions of depolarization factors L_γ for the square and rectangular samples were deduced starting from the corresponding distributions of in-plane NPs semiaxes, extracted from the AFM images of Figs. 6.9b and 6.10b. Due to the rather complex functional dependence of L from the ellipsoids semiaxes [cfr. Eq. (2.43)], the original lognormal shape of the distribution is likely distorted, resulting in a complex distribution of L. For the sake of the present analysis, therefore, the L_γ spread was assumed to obey a normal dispersion for all the ellipsoid axes. The standard deviations σ_L were deduced by converting the spread σ_a of the NPs axes by means

Table 6.1 Morphological parameters deduced from AFM analysis for the square and rectangular samples, employed to calculate the reflectivity spectra reported in Figs. 6.11 and 6.12

Sample	(a_x, a_y, a_z) (nm)	Λ (nm)	d_y (nm)
Square	(11.5, 17, 8)	41	41
Rectangular	(8.3, 8.3, 5)	30	20

For each sample, the ellipsoid semi-axes (a_x, a_y, a_z) are listed, along with the experimental values for Λ and d_y

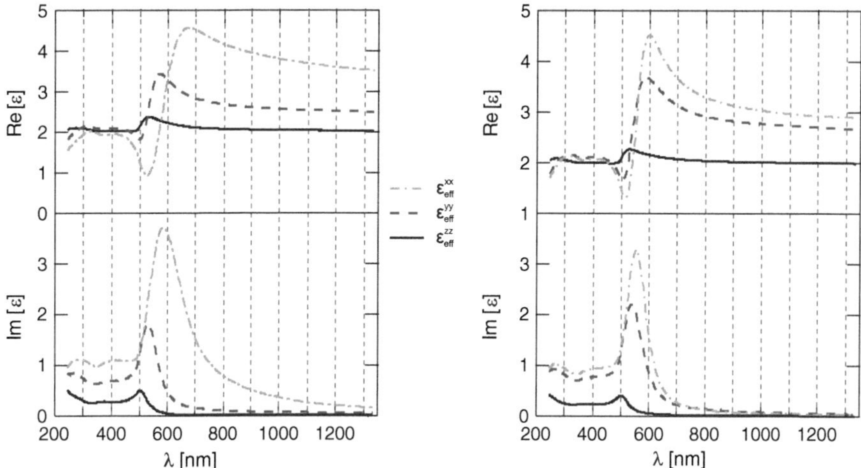

Fig. 6.13 Real and imaginary parts of the computed principal components of the effective dielectric tensor for the square (*left panel*) and rectangular (*right panel*) samples

of Eq. (2.43), i.e. $\sigma_L = [L_\gamma(a_\gamma + \sigma_a) - L_\gamma(a_\gamma - \sigma_a)]/2$. We found $\sigma_L \approx 0.05$ for the rectangular sample and $\sigma_L = 0.06$ for the square case.

Although several strong assumptions have been made, the matching between the experimental data and the theoretical curves is quite good, as can be seen from Figs. 6.11 and 6.12, especially considering the non-trivial experimental geometry addressed (oblique reflectivity) and the achievement of a quantitative agreement between model calculations and data.

The computed effective dielectric functions $\varepsilon_{\text{eff}}^{\xi\xi}$, provided by the model in the two cases, are shown in Fig. 6.13. We can see that, in the proximity of the LSP resonances, ε_{eff} is very similar to the dielectric constant of a Lorentz oscillator (cfr. Fig. 2.3): the imaginary part of ε_{eff} is peaked at the resonance, while the real part exhibits first a minimum and then a maximum crossing the resonance at increasing wavelengths. All the three principal components of the effective dielectric tensor follow the same trend as a function of the wavelength, with variations in the position, width and intensity of the resonances due to the anisotropic NP shape or arrangement. In particular, we notice that the dielectric function $\varepsilon_{\text{eff}}^{zz}$ corresponding to the LSP N mode is considerably smaller than the other ones, in accordance with the fact that such mode is not observed in the R_P reflectivity spectra. We can also qualitatively explain the different positions of the resonances when observed in reflection or transmission (cfr. Figs. 5.9 and 5.7b) in terms of the effective dielectric constant: transmissivity is mostly determined by the absorption of light inside the NPs layer, thus it is mainly related to the imaginary part of ε_{eff}; in contrast, the reflected light is generated by the EM radiation emitted from the induced dipoles, so it more related to the real part of ε_{eff}, whose maximum is at a wavelength slightly longer than the peak of the imaginary part. It should therefore be expected for the LSP peaks observed in reflection to be red-shifted with respect to the corresponding minima of transmissivity.

In order to better clarify the effect of the NPs shape dispersion, the R_S and R_P spectra calculated in absence of L spread are also reported in Figs. 6.11 and 6.12 as the thin dashed lines. Comparing these curves to the ones calculated for $\sigma_L > 0$, we can see that, introducing a dispersion of depolarization factors, both the L and T LSP modes redshift, and the resonances become weaker and broader. These effects are especially pronounced for the L mode of the square sample, i.e. for elongated particles along the major axis, while the corresponding T LSP peak is only slightly modified; for the rectangular sample the finite dispersion effects are instead equally observed for all the involved plasmon modes.

Discussion

In order to fully exploit the possibility of engineering the plasmonic response of the self-organized arrays, the intrinsic single particles properties and the effects of the dipolar coupling on the LSP resonances must be clearly highlighted. In particular, we are interested in understanding how the parameters of fabrication of our samples (substrate temperatures, thickness of deposited gold) affect the LSP, and the relative weights of the intrinsic and collective effects on the optical anisotropy. Calculating the evolution of the LSP peak wavelength as a function of increasing array dimensionality, i.e. gradually introducing mutual interactions in the system along and across the ripples, provides a particularly simple way to achieve this understanding.

We therefore computed the optical response of the square and rectangular samples for three steps of increasing dimensionality: no EM coupling between NPs (0D, equivalent to isolated NPs), EM coupling between NPs allowed only within the *single* chains (1D, where a "chain" is defined as being oriented along the LiF ridges), and EM coupling between all NPs in the array (2D). The variation in the (theoretical) position of the LSP will provide interesting clues about the mechanism of EM coupling in the system.

In Figs. 6.14 and 6.15 we report the calculated R_S, in parallel (panels (a)) and perpendicular (panels (b)) optical configurations (cfr. Fig. 5.5), as a function of the system dimensionality for the square and the rectangular case, respectively; the NPs arrangements sketched in Figs. 6.11e and 6.12e were employed for the calculations. In panels (c) of Figs. 6.14 and 6.15 we also summarize the values of the LSP position

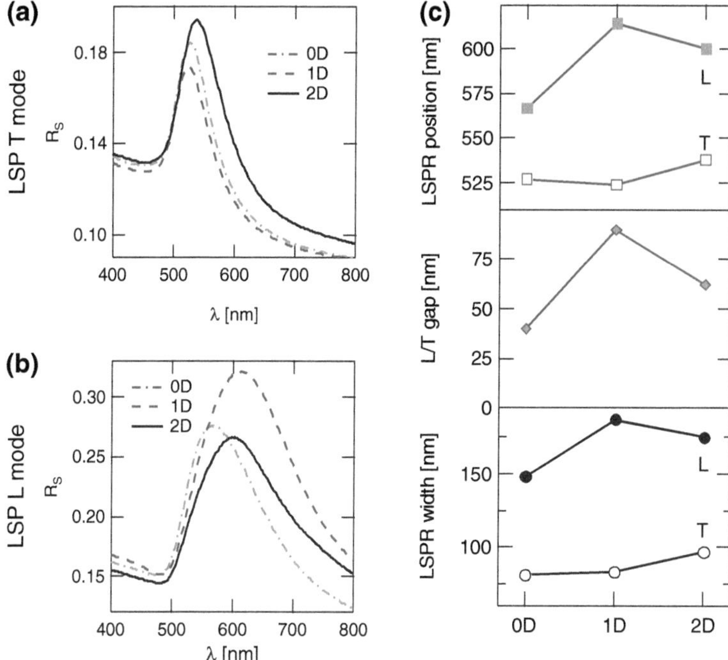

Fig. 6.14 Panels **a**, **b** R_S reflectivity spectra computed as a function of the system dimensionality for the "square" sample, in parallel (**a**) and perpendicular (**b**) optical configurations. Panel **c** wavelength position and linewidth of longitudinal (*solid markers*) and transverse (*open markers*) LSP resonances and their splitting (*diamonds*) as a function of the system dimensionality, deduced from the R_S spectra in panels (**a**, **b**)

($\lambda_{T,P}^{S}$, solid and open squares) and width ($\Gamma_{T,P}^{S}$, solid and open circles) as a function of the dimensionality, along with the L/T modes splitting (solid diamonds).

From a quick inspection of the graphs, it can be clearly seen that the LSPs wavelength and their corresponding width evolve as a function of the system's dimensionality. As a general trend, the variations of both λ^S and Γ^S are not monotonous, and different for the T and L modes.

Let us first discuss in depth the square case: there, the particles are equally spaced and ellipsoidal, with the longest axis parallel to the LiF ridges.

0D Due to the anisotropic shape of the particles, the 0D R_S spectra (red curves in Fig. 6.14a, b) already exhibit intrinsically non-degenerate peaks, located at $\lambda_L^0 = 567$ nm and $\lambda_T^0 = 527$ nm, for exciting field parallel and transverse to the LiF ridges, respectively. The corresponding L/T splitting is about 40 nm, while the plasmon linewidths are $\Gamma_L^0 = 148$ nm and $\Gamma_L^0 = 81$ nm.

1D Introducing the EM coupling between NPs belonging to the same chain leads to the appearance of an induced field $\mathbf{E}_{\text{ind}}$ radiated by the neighbouring NPs (see Fig. 6.16, top panels). When the electric field is transverse to the chain,

$\mathbf{E}_{\text{ind}}$ counteracts the external field (red arrow in Fig. 6.16a), thereby blue shifting the resonance; in contrast, for longitudinal excitation, the induced field adds up to the external field (Fig. 6.16b), with a corresponding red shift of the LSP peak (cfr. Sect. 2.3.4) [27–32, 20, 33, 21, 22]. Moreover, due to the retardation effects of the interactions, the LSP modes become also broader. In our case, the onset of the dipolar coupling would shift the L and T resonance peaks (blue curves in Fig. 6.14a, b) to $\lambda_L^1 = 614\,\text{nm}$ and $\lambda_T^1 = 524\,\text{nm}$, respectively, more than doubling the original 0D splitting (90 nm); correspondingly, the LSP widths become $\Gamma_L^0 = 187\,\text{nm}$ and $\Gamma_L^0 = 83\,\text{nm}$.

2D In 2D, the local field acting on each particle acquires dipolar contributions $\mathbf{E}_{\text{ind}}^*$ also from NPs belonging to neighbouring chains, indicated in blue in Fig. 6.16c, d. For exciting fields both longitudinal and transverse to the ripples, $\mathbf{E}_{\text{ind}}^*$ systematically counteracts the dipolar field $\mathbf{E}_{\text{ind}}$ generated by NPs from the same chain (red arrows), thus reducing the LSP shifts observed in the 1D case. Accordingly, the calculated L mode blue-shifts back to $\lambda_L = 600\,\text{nm}$ while the T mode red-shifts to $\lambda_T = 538\,\text{nm}$ (black curves in Fig. 6.14a, b), yielding a L/T splitting of 62 nm, in agreement with the experimental observations.

For the "rectangular" sample (Fig. 6.15) the situation is generally similar, but with some extremely interesting changes.

0D The in-plane spherical aspect of the NPs yields degenerate single-particle L and T LSP modes, located at $\lambda_L^0 = \lambda_T^0 = 538\,\text{nm}$ and having a linewidth $\Gamma_L^0 = \Gamma_T^0 = 77\,\text{nm}$, i.e. no L/T splitting in the 0D case.

1D The degeneracy is lifted in 1D, due to the presence of the dipolar field $\mathbf{E}_{\text{ind}}$; here a 30 nm splitting is observed, originating from the simultaneous redshift of L excitations to $\lambda_L^1 = 561\,\text{nm}$ and the blueshift of the T peak to $\lambda_T^1 = 532\,\text{nm}$. The width of the L resonances increases to $\Gamma_L^1 = 94\,\text{nm}$, Γ_T remains unchanged.

2D In 2D, the L mode slightly blue shift at $\lambda_T = 557\,\text{nm}$, while the T mode red-shifts to $\lambda_L = 544\,\text{nm}$, finally yielding a L/T splitting of merely 18 nm, significantly lower than the one observed for the square samples. Correspondingly, the linewidths of the resonances increase up to $\Gamma_T^2 = 103\,\text{nm}$ for the T mode and slightly decrease to $\Gamma_L^2 = 92\,\text{nm}$ for the L mode.

Interestingly, we notice that although the NPs lay on a square grid, the L and T peak positions (and their corresponding splitting) in the 2D case are not equivalent to their 0D value. This happens because the anisotropic polarizability of the individual Au inclusions induces a correspondingly anisotropic radiated EM field [33, 34], that further reinforces the intrinsic system birefringence. This is schematically represented in panels (c) and (d) of Fig. 6.16. When the external field is applied along the ripples, i.e. parallel to the major axis of the ellipsoids, the induced dipoles are weaker than for transversal excitations. Correspondingly, the irradiated fields $\mathbf{E}_{\text{ind}}$ and $\mathbf{E}_{\text{ind}}^*$ depend on the direction of the external field, so, despite the square symmetry of the lattice, the collective L and T LSP modes are not degenerate. Thus, the collective optical anisotropy of the square samples arises from a superposition of single-NP

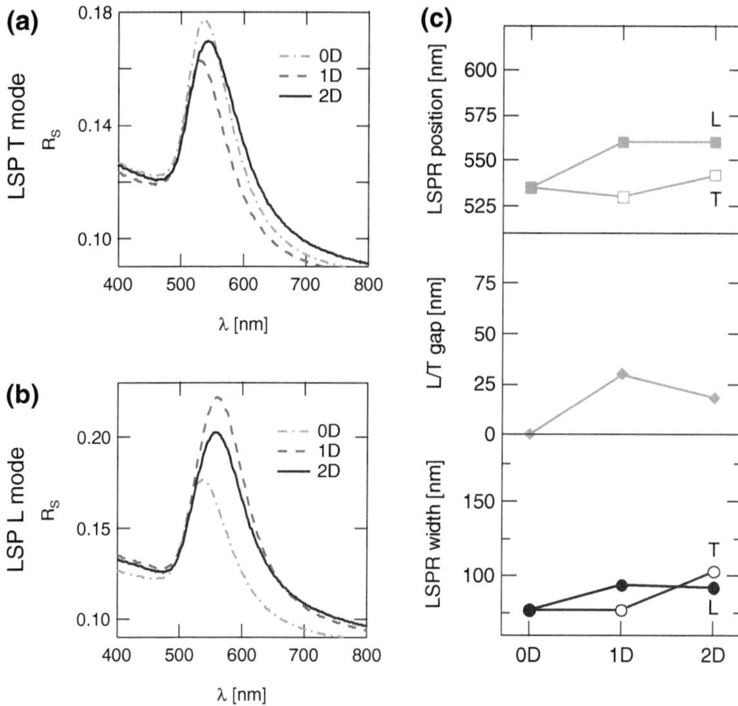

Fig. 6.15 Panels **a**, **b** R_S reflectivity spectra computed as a function of the system dimensionality for the "rectangular" sample, in parallel (**a**) and perpendicular (**b**) optical configurations. Panel **c** wavelength position and linewidth of longitudinal (*solid markers*) and transverse (*open markers*) LSP resonances and their splitting (*diamonds*) as a function of the system dimensionality, deduced from the R_S spectra in panels (**a**, **b**)

effects (i.e. the splitting in 0D) and of anisotropic EM collective coupling, the latter having the net effect of enhancing the original L/T splitting.

In the rectangular case, in contrast, the circular in-plane aspect of the NPs yields an isotropic polarizability of the inclusions, hence no intrinsic 0D L/T splitting and no subsequent anisotropy in the EM radiation. The only source of anisotropic plasmonic response is therefore the rectangular symmetry of the array, which induces the appearance of different dipolar fields $\mathbf{E}_{\mathrm{ind}}$ and $\mathbf{E}^*_{\mathrm{ind}}$ for longitudinal or transverse excitations, due to the not equivalent NPs spacings within the array; these effects are however somewhat smaller as compared to the square sample.

We can therefore conclude that the tuning of the plasmon response of self-organized 2D arrays appears to be more flexibly and successfully performed when starting from anisotropic elementary building blocks, for which intrinsic and collective properties effectively reinforce each other in determining the overall LSP characteristics. Ellipsoidal nanoparticles with different aspect ratios provide the most efficient way to tune the position of the LSP resonances, even within isotropic envi-

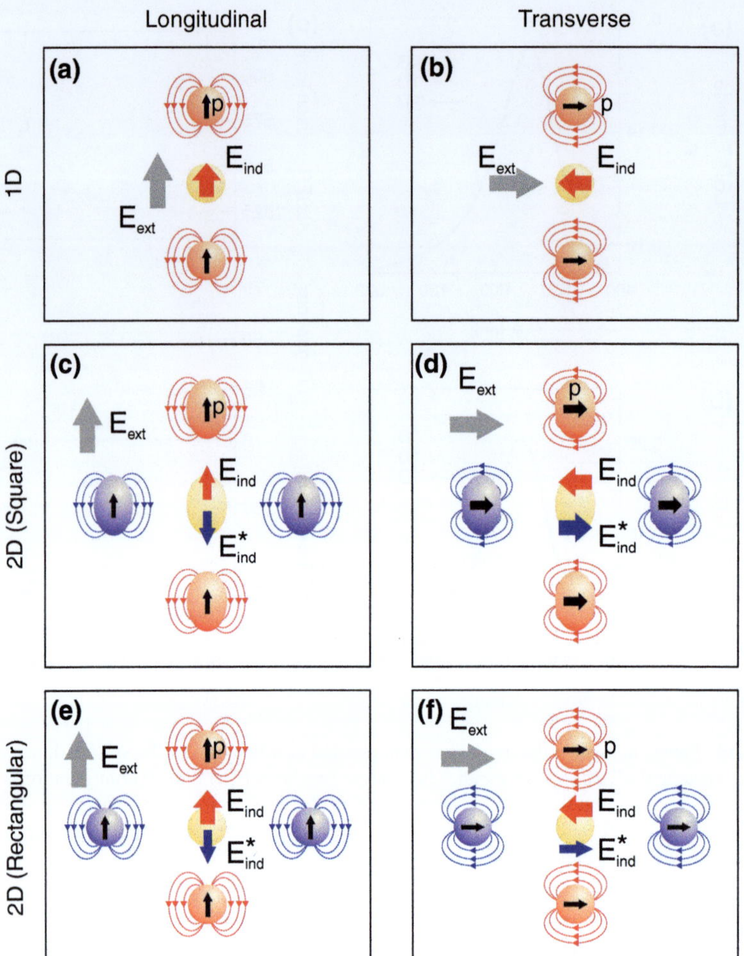

Fig. 6.16 Schematic representation of the dipolar coupling between neighbouring NPs in 1D (**a**, **b**) and 2D (**c**, **d**: square sample; **e**, **f**: rectangular sample), under transverse (*left panels*) and longitudinal (*right panels*) excitation. *Red spheres/arrows* particles belonging to the same chain, and corresponding dipolar field $\mathbf{E}_{ind}$; *blue spheres/arrows* particles and dipolar field $\mathbf{E}^*_{ind}$ from adjacent ripples

ronments, at the cost of broader resonances. For spherical particles, instead, the LSP peaks are very sharp, however the only source of tunability of the plasmonic response is the arrangement of the NPs, which has limited effects due to the partial cancellation of the EM interactions within the array.

References

1. H. H. Li. Refractive index of alkali halides and its wavelength and temperature derivatives. *J. Phys. Chem. Ref. Data*, 5:329, 1976.
2. H. Fujiwara. *Spectroscopic Ellipsometry. Principles and Applications*. Wiley, 2007.
3. G. Baldacchini, O. Goncharova, V. S. Kalinov, R. M. Montereali, E. Nichelatti, A. Vincenti, and A. P. Voitovich. Optical properties of coloured lif crystals with given content of oxygen, hydroxyl and metal impurities. Phys. Stat. Sol. (c), 4:744, 2007.
4. G. Baldacchini, O. Goncharova, V. S. Kalinov, R. M. Montereali, A. Vincenti, and A. P. Voitovich. Thermal transformation of colour centres in lif crystals with given content of oxygen, hydroxyl and metal ions. Phys. Stat. Sol. (c), 4:1134, 2007.
5. Bruce T. Draine and Piotr J. Flatau. Discrete-dipole approximation for scattering calculations. *J. Opt. Soc. Am. A*, 11:1491, 1994.
6. B. T. Draine. The discrete-dipole approximation and its application to interstellar graphite grains. *Astr. J.*, 333:848, 1988.
7. P. C. Waterman. Matrix methods in potential theory and electromagnetic scattering. *J. App. Phys.*, 50:4550, 1979.
8. Michael I. Mishchenko, Larry D. Travis, and Daniel W. Mackowski. T-matrix computations of light scattering by nonspherical particles: A review. *J. Quant Spectrosc. Ra.*, 55:535, 1996.
9. Jian-Ming Jin. *The Finite Element Method in Electromagnetics, 2nd Edition*. Wiley-IEEE Press, 2002.
10. L. Gao, K. W. Yu, Z. Y. Li, and Bambi Hu. Effective nonlinear optical properties of metal-dielectric composite media with shape distribution. *Phys. Rev. E*, 64:036615, 2001.
11. J.D. Jackson. *Classical Electrodynamics*. John Wiley & Sons, 1999.
12. Bohren C.F.; Huffman D.R. *Absorption and scattering of light by small particles*. Wiley, 1998.
13. Rubén G. Barrera, Pedro Villaseñor González, W. Luis Mochán, and Guillermo Monsivais. Effective dielectric response of polydispersed composites. *Phys. Rev. B*, 41:7370, 1990.
14. Rubén G. Barrera, Marcelo del Castillo-Mussot, Guillermo Monsivais, Pedro Villaseor, and W. Luis Mochán. Optical properties of two-dimensional disordered systems on a substrate. *Phys. Rev. B*, 43:13819, 1991.
15. Rubén G. Barrera, Jairo Giraldo, and W. Luis Mochán. Effective dielectric response of a composite with aligned spheroidal inclusions. *Phys. Rev. B*, 47:8528, 1993.
16. Bashara N.M. Azzam R.M.A. *Ellipsometry and Polarized Light*. North Holland, 1988.
17. Stenzel O. *The Physics of Thin Film Optical Spectra: An Introduction*. Springer, 2010.
18. Wolf E. Born M. *Principles of Optics*. Cambridge University Press, 1999.
19. W. Rechberger, A. Hohenau, A. Leitner, J. R. Krenn, B. Lamprecht, and F. R. Aussenegg. Optical properties of two interacting gold nanoparticles. *Opt. Comm.*, 220:137, 2003.
20. Prashant K. Jain, Wenyu Huang, and Mostafa A. El-Sayed. On the universal scaling behavior of the distance decay of plasmon coupling in metal nanoparticle pairs: A plasmon ruler equation. *Nano Letters*, 7:2080, 2007.
21. Christopher Tabor, Raghunath Murali, Mahmoud Mahmoud, and Mostafa A. El-Sayed. On the use of plasmonic nanoparticle pairs as a plasmon ruler: The dependence of the near-field dipole plasmon coupling on nanoparticle size and shape. *J. Phys. Chem. A*, 113:1946, 2009.
22. Prashant K. Jain and Mostafa A. El-Sayed. Plasmonic coupling in noble metal nanostructures. *Chem. Phys. Lett.*, 487:153, 2010.
23. H. Hövel, S. Fritz, A. Hilger, U. Kreibig, and M. Vollmer. Width of cluster plasmon resonances: bulk dielectric functions and chemical interface damping. *Phys. Rev. B*, 48:18178, 1993.
24. E. Stefan Kooij, Herbert Wormeester, E. A. Martijn Brouwer, Esther van Vroonhoven, Arend van Silfhout, and Bene Poelsema. Optical characterization of thin colloidal gold films by spectroscopic ellipsometry. *Langmuir*, 18:4401, 2002.
25. F. Bisio, M. Palombo, M. Prato, O. Cavalleri, E. Barborini, S. Vinati, M. Franchi, L. Mattera, and M. Canepa. Optical properties of cluster-assembled nanoporous gold films. *Phys. Rev. B*, 80:205428, 2009.

26. Stephan Link and Mostafa A. El-Sayed. Size and temperature dependence of the plasmon absorption of colloidal gold nanoparticles. *J. Phys. Chem. B*, 103:4212, 1999.
27. M. Quinten, A. Leitner, J. R. Krenn, and F. R. Aussenegg. Electromagnetic energy transport via linear chains of silver nanoparticles. *Opt. Lett.*, 23:1331, 1998.
28. Mark L. Brongersma, John W. Hartman, and Harry A. Atwater. Electromagnetic energy transfer and switching in nanoparticle chain arrays below the diffraction limit. *Phys. Rev. B*, 62:R16356, 2000.
29. Stefan A. Maier, Pieter G. Kik, and Harry A. Atwater. Observation of coupled plasmon-polariton modes in Au nanoparticle chain waveguides of different lengths: Estimation of waveguide loss. *App. Phys. Lett.*, 81:1714, 2002.
30. L. L. Zhao, K. L. Kelly, and G. C. Schatz. The extinction spectra of silver nanoparticle arrays: Influence of array structure on plasmon resonance wavelength and width. *J. Phys. Chem. B*, 107:7343, 2003.
31. Christy L. Haynes, Adam D. McFarland, LinLin Zhao, Richard P. Van Duyne, George C. Schatz, Linda Gunnarsson, Juris Prikulis, Bengt Kasemo, and Mikael Käll. Nanoparticle optics: The importance of radiative dipole coupling in two-dimensional nanoparticle arrays. *J. Phys. Chem. B*, 107:7337, 2003.
32. Erin M. Hicks, Shengli Zou, George C. Schatz, Kenneth G. Spears, Richard P. Van Duyne, Linda Gunnarsson, Tomas Rindzevicius, Bengt Kasemo, and Mikael Käll. Controlling plasmon line shapes through diffractive coupling in linear arrays of cylindrical nanoparticles fabricated by electron beam lithography. *Nano Lett.*, 5:1065, 2005.
33. Christopher Tabor, Desiree Van Haute, and Mostafa A. El-Sayed. Effect of orientation on plasmonic coupling between gold nanorods. *ACS Nano*, 3:3670, 2009.
34. K.-H. Su, Q.-H. Wei, X. Zhang, J. J. Mock, D. R. Smith, and S. Schultz. Interparticle coupling effects on plasmon resonances of nanogold particles. *Nano Letters*, 3:1087, 2003.

Chapter 7
Composite Magnetic-Plasmonic Media Based on Au/LiF Arrays

In this chapter, we report a selected example of fabrication of a 2D array of magnetic nanoparticles, realized employing the array of gold nanoparticles as a template for guiding the deposition of the magnetic clusters. Such a composite system, with both optical and magnetic functionality, would represent the starting point to realize, for example, optically active magneto-plasmonic media, with tunable LSPs controlled by the application of an external magnetic field. For this purpose, we started by investigating the deposition of magnetite (Fe_3O_4) nanoparticles on the Au NPs arrays from a colloidal suspension, and characterizing the morphology and the variations of the optical properties induced by the presence of the magnetic medium.

Solutions of Fe_3O_4 nanoparticles coated with oleic acid (Fe_3O_4/OA) dispersed in hexane were synthesized in collaboration with CSIC/ICMM in Madrid, following established procedures [1–3]; oleic acid was chosen as surfactant in order to prevent the aggregation and the precipitation of the particles. Fe_3O_4 nanocrystals were prepared by the thermal decomposition of iron (III)-oleate complex in 1-octadecen, in the presence of oleic acid in a relation of 3:1 with respect to the metal complex. The solution was slowly heated under nitrogen atmosphere to $\approx$300 °C, and then aged at that temperature for 240 min while stirred, generating the iron oxide nanocrystals. A representative image of some nanoparticles dried from the colloidal suspension, acquired by transmission electron microscopy (TEM), is reported in Fig. 7.1a (courtesy of ICMM). The image reveals the magnetite cores of the particles, which appear cubic in shape. From a statistical analysis of the NPs sizes, the diameters of the cores d_{core} resulted lognormally distributed, with a mean and a standard deviation of $d_{core} = (13.0 \pm 1.3)$ nm.

The colloidal suspension was characterized by dynamic light scattering (DLS) [4], an experimental technique based on light scattering from small particles, which allows to deduce the *hydrodynamic* size d_{hydro} of the particles, including the length of the surfactant molecules. The statistical distribution of d_{hydro} obtained from DLS is reported in Fig. 7.1b: similarly to TEM analysis, the NPs sizes are lognormally distributed, and the diameters have a mean and standard deviation of $d_{hydro} = (16 \pm 4)$ nm. Comparing this result with the values from TEM analysis, we obtain a mean

L. Anghinolfi, *Self-Organized Arrays of Gold Nanoparticles*, Springer Theses, DOI: 10.1007/978-3-642-30496-5_7, © Springer-Verlag Berlin Heidelberg 2012

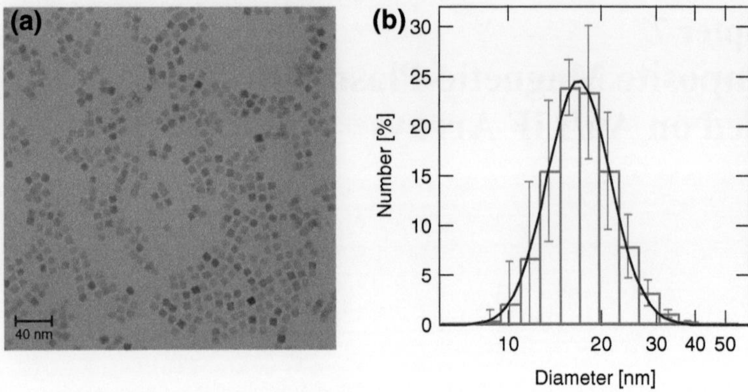

Fig. 7.1 *Left panel* transmission electron microscopy image (*courtesy of ICMM*) of Fe_3O_4 nanoparticles obtained by decomposition of an iron oleate complex in 1-octadecen and oleic acid (see text for details). *Right panel* distribution of the particles hydrodynamic size obtained from dynamic light scattering analysis on a colloidal suspension of Fe_3O_4/AO particles dispersed in hexane

length of the surfactant molecules of ≈ 1.5 nm, comparable with the typical length of oleic acid molecules of 1.8 nm [5]. The distributions of sizes also show that the suspension is monodisperse, and that the tendency of the NPs to form aggregates is negligible.

The template for the deposition of the Fe_3O_4/OA nanoparticles was prepared according to the procedure described in the previous chapters. First, a nanopatterned LiF(110) substrate was grown following the homoepitaxial deposition of $t_{LiF} \approx 250$ nm LiF at $T = 300\,°C$, which induced the formation of a well-developed ripple structure with a mean periodicity $\Lambda \approx 35$ nm; an AFM image of the surface is shown in Fig. 7.2a. Then, the 2D array of gold nanoparticles was obtained by grazing deposition of $t_{Au} \approx 5$ nm gold at $T = 100\,°C$ and subsequent annealing at $T = 400\,°C$. We report in Fig. 7.2b a representative AFM image of the array, where we can distinguish parallel chains of slightly elongated nanoparticles, oriented along the LiF ridges, and with mean in-plane semi-axes of $a_x \approx 11$ nm and $a_y \approx 14$ nm, respectively across and parallel the ripples. The corresponding Fourier spectrum, shown in the inset of the figure, confirms the ordered arrangement of the particles: the sharp intensity peak confined in the $[1\bar{1}0]$ direction, and rapidly decaying in the [001] direction, shows that the periodicity of the underlying ripple structure has been preserved within the Au NPs array, while the broader feature slowing decaying in all directions, and slightly stretched in the $[1\bar{1}0]$ direction, reflects the in-plane anisotropic shape of the particles.

The deposition of the Fe_3O_4/OA NPs has been performed by immersing the Au/LiF substrate in a colloidal suspension of NPs dispersed in hexane with a concentration of $\approx 6\,\mu g$ Fe per ml, for a period of time of about 5 min; then, the sample has been washed in pure hexane, in order to remove the particles weakly stuck to the surface, and dried under nitrogen flux. An AFM image measured after the deposition

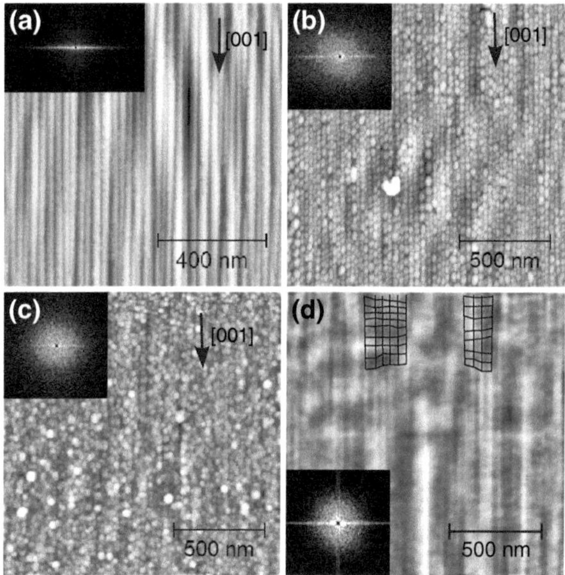

Fig. 7.2 *Top panels* AFM images of a nanopatterned LiF(110) substrate, after the deposition of $t_{LiF} \approx 250$ nm LiF at $T = 300\,°C$ (panel (**a**)), and the grazing deposition of $t_{Au} \approx 5$ nm gold at $T = 100\,°C$ and subsequent annealing at $T = 400\,°C$ (panel (**b**)). Panel (**c**): AFM image of the same sample after 5' immersion in solution of Fe_3O_4/OA nanoparticles dispersed in hexane ($\approx 6\mu$g Fe per ml). Panel (**d**): cross correlation function of the images in panels (**b**) and (**c**); the superimposed grid is a guide to the eye for the positions of the maxima of the correlation. The Fourier spectral densities for each image are shown in the insets of the panels

of the magnetic NPs is reported in Fig. 7.2c. The surface appears uniformly covered with Fe_3O_4/OA nanoparticles, and no Au particle could be apparently singled out. From the sole inspection of the AFM image we cannot determine the exact thickness of the magnetic NP layer. However earlier depositions on silicon substrates (not reported here) showed that the Fe_3O_4/OA particles did not form more than a monolayer even after prolonged immersions of several hours in solution; we can therefore expect that also in this case a single monolayer of magnetic nanoparticles was deposited.

Looking at the AFM image of Fig. 7.2c, the particles appear spherical, their cubic shape being smoothed by the convolution effects of the AFM tip, and the preferential orientation imposed by the underlying ripple structure can still be recognized. In fact, the characteristic peak of intensity confined in the $[1\bar{1}0]$ direction, fingerprint of the LiF ripples periodicity, is still present in the Fourier spectrum (inset of the figure), although less pronounced than on the bare Au NPs array. Compared to Fig. 7.2b for the Au NPs, it is less intuitive in Fig. 7.2c to identify the Fe_3O_4/OA NPs arrangement; furthermore, since the typical spacing between the gold NPs is ≈ 33 nm, against a mean Fe_3O_4/OA NPs size of ≈ 16 nm, there is in principle enough space to accommodate more than a single magnetic NP per Au NP, therefore the arrangement of the Au NPs can hardly be exactly reproduced in the Fe_3O_4/OA layer. Nevertheless, we can try to pick out a correlation between the Au and the Fe_3O_4/OA NPs arrays by computing the cross correlation function of the two previous AFM images; this is shown in Fig. 7.2d. The cross correlation reveals a pattern of maxima and minima, which form an almost square grid with typical spacings of $\approx$40 nm, comparable to the ripple periodicity and the typical spacings of the Au NPs. The pattern is particularly emphasized by the corresponding Fourier spectral density (inset of Fig. 7.2d), which is characterized by two sharp peaks confined along the $[1\bar{1}0]$ and the $[001]$ directions, indicating the presence of periodic features both along and across the ripples. These observations suggest that, even if the Fe_3O_4/OA NPs do not follow the Au NPs arrangement with a one-to-one correspondence, still the Au NPs provide preferential sites for the adsorption of the magnetic NPs, so that their arrangement can be effectively reproduced by groups of few Fe_3O_4/OA NPs.

Similarly to the previous cases, the optical response of the sample was optically characterized by means of s- and p-polarized light reflectivity, with the plane of incidence either parallel or perpendicular to the LiF ripples. The R_S and R_P spectra acquired at an incidence angle of $\theta = 50°$, before and after the Fe_3O_4/OA deposition, are reported in Fig. 7.3.

The presence of the magnetite layer affects the LSPs by systematically redshifting the resonances and increasing their width; the overall intensity of the reflectivity is also increased. Before the deposition (red curves in Fig. 7.3), the longitudinal LSP mode is found at $\lambda_L = 620$ nm, with a width of $\Gamma_L = 170$ nm; the transverse mode is instead located at $\lambda_T = 570$ nm and has a width of $\Gamma_T = 125$ nm. Following the deposition (black curves in Fig. 7.3), both modes redshift by few tens of nm, the L mode at $\lambda_L^* = 645$ nm and the T mode at $\lambda_T^* = 585$ nm, while the corresponding widths become $\Gamma_L^* = 215$ nm and $\Gamma_T^* = 145$ nm, respectively.

This trend is in accordance with the qualitative considerations in Sect. 2.3.3, where we discussed the dependence of the dielectric environment on the LSP resonances. In particular, the simultaneous redshift of the resonances can be explained in terms of the Fröhlich condition $[\varepsilon_m^{L,T}(\lambda_{L,T})] = -2\,\varepsilon_h$ (see Eq. 2.45), where ε_m, $\varepsilon_h^{L,T}$ and $\lambda_{L,T}$ are the dielectric constant of the individual Au NPs and of the host and the position of the resonances, respectively. ε_m is shown in Fig. 6.6b, and its real part has a negative slope for increasing wavelengths; in Fig. 7.4 we report instead the dielectric constant of bulk magnetite, extracted from [6]. In analogy with the model of the previous chapter, in first approximation the dielectric constant ε_h can be expressed as a linear combination of the dielectric constants of the materials surrounding the Au NPs; therefore, following the deposition of the Fe_3O_4/OA NPs, ε_h is expected to increase, because the real part of the dielectric constant of magnetite reads about $\varepsilon_1 = 5$ in correspondence of the gold nanoparticles LSP resonances. Then, according to the Fröhlich condition, if ε_h increases and ε_m has a negative slope as a function of the wavelength, the LSP resonances of the gold NPs must red shift when embedded in the Fe_3O_4/OA layer.

In order to quantitatively evaluate the effects of the Fe_3O_4/OA NPs on the plasmonic response of the Au NPs array, we can apply the theoretical framework

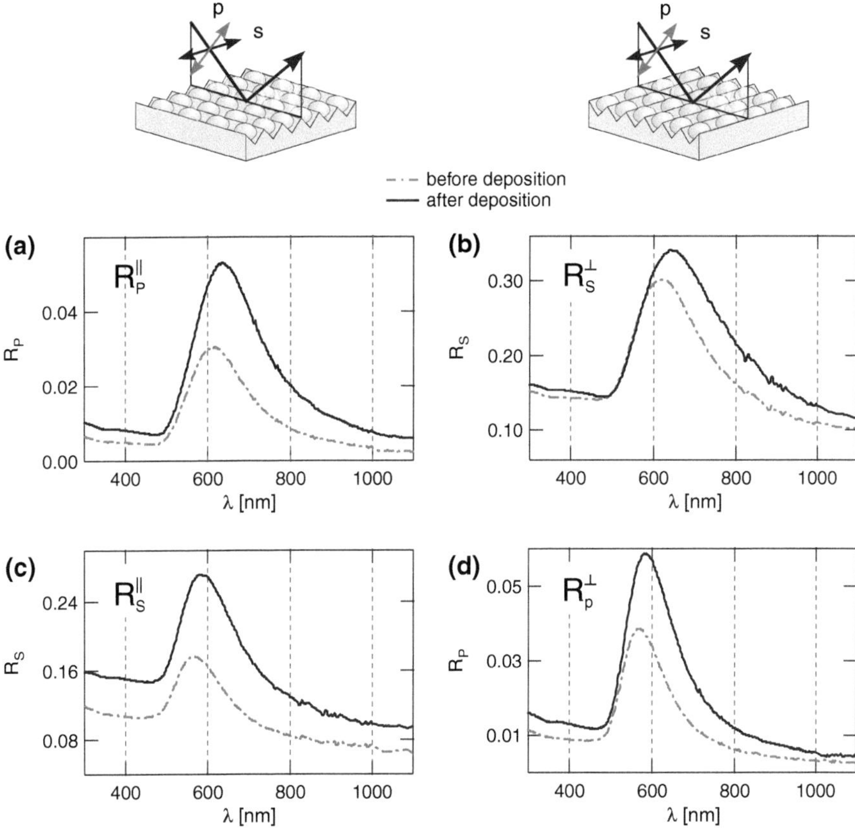

Fig. 7.3 Reflectivities, at $\theta = 50°$, measured before (*red curves*) and after (*black curves*) the deposition of $\approx$1 monolayer of Fe_3O_4OA NPs on 2D arrays of gold NPs (see Fig. 7.2). Plane of incidence either parallel (*left panels*) and perpendicular (*right panels*) to the ripples. EM excitation along (panels (**a**), (**b**)) and across (panels (**c**), (**d**)) the ripples

developed in the previous chapter to the current system. In Fig. 7.5a we report the computed R_S spectra for the "bare" Au/LiF nanostructures, compared to the corresponding experimental values, for the L and T LSP modes, at an angle of incidence of $\theta = 50°$. In analogy with before, the morphological parameters employed for the calculations were obtained from the analysis of the AFM data: for the NPs spacings across and along the ripples we used $d_x = d_y = 33$ nm, while the values of the ellipsoids semiaxes were $a_x = 11.2$ nm, $a_y = 14.0$ nm and $a_z = 6.4$ nm.

The corresponding spectra in presence of the Fe_3O_4/OA NPs were computed keeping the same input parameters, except two specific modifications. Considering a single monolayer of deposited NPs, the first one was to increase the thickness of the effective layer by an amount equal to the mean diameter of the NPs, i.e. $\approx$17 nm. The second correction was to rewrite the dielectric constant ε_h of the host including

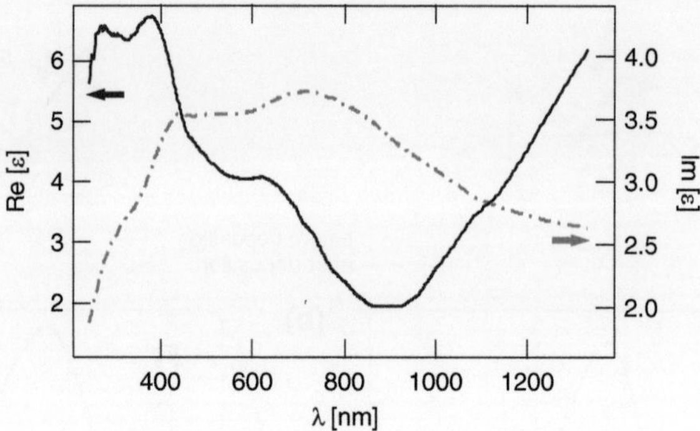

Fig. 7.4 Real (*red line*) and imaginary (*black line*) parts of the dielectric constant of bulk magnetite in the visible and near-IR range, from [6]

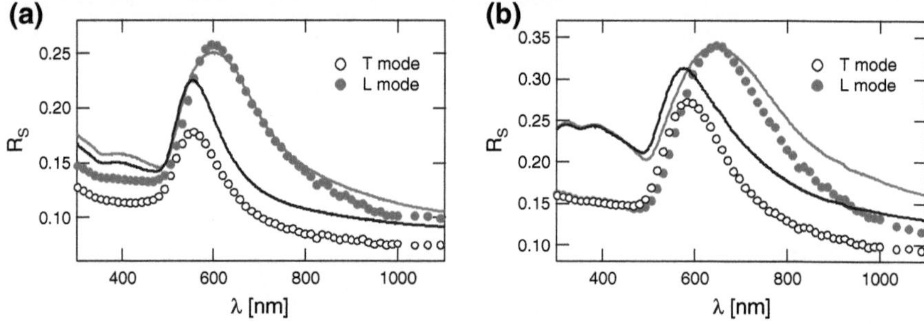

Fig. 7.5 Experimental (*circles*) and computed (*lines*) R_S spectra, with plane of incidence parallel (*black curves*) and perpendicular (*red curves*) the LiF ridges. *Left panel* 2D arrays of gold NPs on nanopatterned LiF(110) ($\Lambda = 35$ nm, $t_{Au} \approx 5$ nm, AFM data in Fig. 7.2a, b). *Right panel* same system after the deposition of a monolayer of Fe$_3$O$_4$/OA NPs (AFM data in Fig. 7.2c)

the contribution of the magnetite NPs; we considered the gold NPs embedded in an effective host composed at 50 % by LiF (with dielectric constant ε_s) and at 50% by a mixing of air and bulk magnetite (with dielectric constant $\varepsilon_{Fe_3O_4}$, assuming that the dielectric constant of Fe$_3$O$_4$ nanoparticles does not differ from the bulk):

$$\varepsilon_h = \frac{1}{2}\varepsilon_s + \frac{1}{2}(f\,\varepsilon_{Fe_3O_4} + (1-f)\varepsilon_{air}) \qquad (7.1)$$

where $\varepsilon_{air} = 1$ and f is the magnetite filling factor. The latter is calculated by simple geometrical considerations: assuming the Fe$_3$O$_4$/OA nanoparticles arranged on a square lattice, and in contact with one another, we find one particle per volume of d_{hydro}^3 (d_{hydro} is the hydrodynamic size), while the volume of the magnetite core is $4\pi/3(d_{core}/2)^3$ (d_{core} is the size of the NPs core); the filling factor f is given by the ratio of the two volumes and, substituting the values obtained from TEM and DLS measurements, results $f \approx 0.3$.

The calculated R_S spectra for the Au NPs array after the Fe_3O_4/OA deposition are reported in Fig. 7.5b. Despite the simplifications in treating the magnetite NPs layer, which probably determined the overestimation of the absolute values of reflectivity, the positions and widths of the LSPs were reproduced in good agreement with the experiments, resulting of $\lambda_L^* = 645\,nm$ and $\Gamma_L^* = 223\,nm$ for the L mode and $\lambda_T^* = 576\,nm$ and $\Gamma_T^* = 145\,nm$ for the T mode. In particular, this also supports the consideration that only one monolayer of Fe_3O_4/OA NPs has been deposited.

In conclusion, we have shown that the self-organized Au/LiF systems could be fruitfully exploited as templates for the fabrication of more complex composite structures, that can at least partly retain the morphological characteristics of the original system and its optical response, adding to these novel functionalities, like magnetic response in this specific case.

References

1. A G Roca, M P Morales, K OGrady, and C J Serna. Structural and magnetic properties of uniform magnetite nanoparticles prepared by high temperature decomposition of organic precursors. *Nanotechnology*, 17:2783, 2006.
2. A.G. Roca, M.P. Morales, and C.J. Serna. Synthesis of monodispersed magnetite particles from different organometallic precursors. *IEEE Trans. Mag.*, 42:3025, 2006.
3. P. Guardia, B. Batlle-Brugal, A.G. Roca, O. Iglesias, M.P. Morales, C.J. Serna, A. Labarta, and X. Batlle. Surfactant effects in magnetite nanoparticles of controlled size. *J. Mag. Mag. Mat.*, 316:e756, 2007.
4. B.J. Berne and R. Pecora. *Dynamic Light Scattering*. Wiley, New York, 1976.
5. Calculated from NIST Chemistry Webbook data (http://webbook.nist.gov/cgi/cbook.cgi?ID=C2027476&Units=SI)
6. W. F. J. Fontijn, P. J. van der Zaag, M. A. C. Devillers, V. A. M. Brabers, and R. Metselaar. Optical and magneto-optical polar Kerr spectra of Fe_3O_4 and Mg^{2+}- or Al^{3+}-substituted Fe_3O_4. *Phys. Rev. B*, 56:5432, 1997.

Chapter 8
Conclusions

In this thesis, we have discussed the fabrication of self-organized 2-dimensional arrays of gold nanoparticles, with independently tunable size, shape and periodic arrangement, and the characterization of their collective plasmonic response.

The arrays were fabricated employing self-organized nanopatterned LiF(110) substrates as templates for guiding the particle formation. First, a regular ridge-and-valley (ripple) structure is spontaneously induced by means of homoepitaxial growth of LiF on LiF(110), with a variable periodicity Λ determined by the substrate temperature during the deposition. Typical values of $\Lambda = 25$–$60\,$nm by varying the temperature in the range between 250 and 450 °C were found.

Then, a thin layer of gold was deposited at incidence of $\theta = 60°$, exploiting the shadow effect of the ripples ridges in order to form disconnected Au "nanowires". Finally, the samples were mildly annealed at $T = 400\,$°C, promoting the dewetting of the gold nanowires into individual nanoparticles.

After the dewetting, the surface of the samples was characterized by parallel chains of gold nanoparticles, regularly spaced according to the periodicity of the underlying ripples. The particles exhibited lognormal size distribution, with a typical standard deviation lower than 5 nm. Interestingly, we found a partial correlation also between the positions of particles belonging to adjacent ripples, and rectangular arrangements of Au NPs could be observed over large areas of the samples.

By tuning the fabrication parameters, we showed the possibility of varying both the single NP mean size and in-plane aspect ratio and their mutual spacings in the arrays. The ripples periodicity Λ determined the NPs mean size and spacing in the direction transversal to the LiF ridges, while the length and spacing along the ripples were related to the amount of deposited gold. Performing depositions under different conditions, we could obtain several 2D arrays of gold NPs with in-plane aspect ratios between 1:1 and 1:2, and typical dimensions in the 20–50 nm range.

The 2D arrays of gold NP were optically investigated by means of spectroscopic ellipsometry, reflectivity and transmissivity, in the visible and near-IR frequency range. The optical response of the arrays exhibited characteristic absorptions and reflectivity peaks, corresponding to the excitation of localized surface plasmons, i.e.

L. Anghinolfi, *Self-Organized Arrays of Gold Nanoparticles*, Springer Theses, 121
DOI: 10.1007/978-3-642-30496-5_8, © Springer-Verlag Berlin Heidelberg 2012

collective oscillations of the electron gas confined inside the metallic NPs. Such resonances critically depended on single-particles parameters, like shape and size, and on extrinsic factors, like the dielectric environment or the inter-particles spacing. We investigated the position and width of the LSPs for different arrangements of the Au NPs, with particular emphasis on the optical anisotropy between the longitudinal LSP mode, excited by an electric field along the ripples, and the transverse LSP mode, excited perpendicular to the ripples.

In order to separately assess the contributions of the intrinsic and collective factors on the plasmonic response of the NP systems, we fabricated and discussed in detail two particular cases of 2D arrays. The first featured a 2D array of in-plane spherical gold particles, having different interparticle spacings along and across the ripples, providing an optical anisotropy mainly due to the anisotropic EM coupling between the particles. The second array accommodated elongated gold nanoparticles arranged on a square grid, providing an optical anisotropy mainly due to the anisotropic single-particle polarizability.

The optical properties of the two classes of samples were described by means of a simple but powerful effective medium model, including interparticle dipolar inter-actions, morphological disorder in the NPs shapes, and substrate effects. Combining experiments with model calculations, we showed the dependence of the plasmonic response on the array dimensionality, and we could separately estimate the effects of the single contributions (intrinsic and collective) on the LSP characteristics. In this way, we found that the plasmonic response of the NPs in the arrays with rectangular symmetry is relatively weakly affected by the interparticle EM dipolar coupling, via a partial cancellation effect of the dipolar fields of neighbouring NPs. Instead, the single-particle optical anisotropy of elongated NPs is further increased by the action of the EM interactions, even for square arrays.

To conclude, we presented a method to fabricate 2D arrays of metallic nanopar-ticles entirely based on self-organization processes. The procedure developed here could be applied for the cheap and easy fabrication of flexibly-tunable plasmonic sup-ports for applications in which the high-degree of order achievable by lithographic methods [1–9] is not strictly required, like substrates for LSP-enhanced optical spec-troscopies or LSP-based sensing.

The method has been applied to the case of gold NPs, but it is very general, suggesting facile extension to other materials than gold, for which similar results are expected. In this respect, it can represent an advantageous solution over, for example, the chemical synthesis of nanoparticles, which instead requires specific procedures for each material.

The possibility to employ different materials gives the opportunity to address various spectral regions; in fact, by using metals like Cu, Au, Ag or Al, the position of the LSP resonances can be tuned from the visible to the UV range, and by realizing arrays of metals alloys NPs it could be even possible to fine-tune the optical range of the device.

The same methodology can also be applied to systems with different function-alities than plasmonic. For example nanopatterned NaCl(110) substrates have been employed for the fabrication of ferromagnetic iron nanowires [10–12], iron nanodots and undulating films [12], and arrays of cobalt nanoparticles [13].

Finally, we showed how the 2D arrays of metallic nanoparticles can be employed as base for the realization of composite devices. For example, an optically active device could be obtained by depositing a monolayer of magnetic nanoparticles, where the plasmonic response of the metallic NPs are tuned by the application of an external magnetic field.

References

1. K.-H. Su, Q.-H. Wei, X. Zhang, J. J. Mock, D. R. Smith, and S. Schultz. Interparticle coupling effects on plasmon resonances of nanogold particles. *Nano Letters*, 3:1087, 2003
2. W. Rechberger, A. Hohenau, A. Leitner, J. R. Krenn, B. Lamprecht, and F. R. Aussenegg. Optical properties of two interacting gold nanoparticles. *Opt. Comm.*, 220:137, 2003
3. Prashant K. Jain, Wenyu Huang, and Mostafa A. El-Sayed. On the universal scaling behavior of the distance decay of plasmon coupling in metal nanoparticle pairs: A plasmon ruler equation. *Nano Letters*, 7:2080, 2007
4. Christopher Tabor, Desiree Van Haute, and Mostafa A. El-Sayed. Effect of orientation on plasmonic coupling between gold nanorods. *ACS Nano*, 3:3670, 2009
5. Christopher Tabor, Raghunath Murali, Mahmoud Mahmoud, and Mostafa A. El-Sayed. On the use of plasmonic nanoparticle pairs as a plasmon ruler: The dependence of the near-field dipole plasmon coupling on nanoparticle size and shape. *J. Phys. Chem. A*, 113:1946, 2009
6. Prashant K. Jain and Mostafa A. El-Sayed. Plasmonic coupling in noble metal nanostructures. *Chem. Phys. Lett.*, 487:153, 2010
7. Christy L. Haynes, Adam D. McFarland, LinLin Zhao, Richard P. Van Duyne, George C. Schatz, Linda Gunnarsson, Juris Prikulis, Bengt Kasemo, and Mikael Käll. Nanoparticle optics: The importance of radiative dipole coupling in two-dimensional nanoparticle arrays. *J. Phys. Chem. B*, 107:7337, 2003
8. Mark L. Brongersma, John W. Hartman, and Harry A. Atwater. Electromagnetic energy transfer and switching in nanoparticle chain arrays below the diffraction limit. *Phys. Rev. B*, 62:R16356, 2000
9. Stefan A. Maier, Pieter G. Kik, and Harry A. Atwater. Observation of coupled plasmon-polariton modes in Au nanoparticle chain waveguides of different lengths: Estimation of waveguide loss. *App. Phys. Lett.*, 81:1714, 2002
10. A. Sugawara, D. Streblechenko, M. McCartney, and M.R. Scheinfein. Magnetic coupling in self-organized narrow-spaced Fe nanowire arrays. *IEEE Trans. Mag.*, 34:1081, 1998
11. Akira Sugawara, T. Coyle, G. G. Hembree, and M. R. Scheinfein. Self-organized Fe nanowire arrays prepared by shadow deposition on NaCl(110) templates. *App. Phys. Lett.*, 70:1043, 1997
12. Akira Sugawara, G. G. Hembree, and M. R. Scheinfein. Self-organized mesoscopic magnetic structures. *J. App. Phys.*, 82:5662, 1997
13. A. Sugawara. Quasi-one-dimensional cobalt particle arrays embedded in 5 nm-wide gold nanowires. *IEEE Trans. Mag.*, 37:2123, 2001